Climate Actions

"The climate movement is now old and robust enough to be studied–which in and of itself is good news. And the results of those inquiries are of great use in the task ahead, which is building this movement big enough and fast enough to catch up with the physics of global warming. This book will help make our labors more effective!"

—Bill McKibben, *founder, 350.org, and Schumann Distinguished Scholar in Environmental Studies, Middlebury College, USA*

"Addressing climate change will not be achieved through business as usual. Citizen action is absolutely vital to prod, push and directly institute changes. To learn how citizen action might be improved, Laurence Delina brings to bear insights from social theory combined with reports from numerous action groups. His classification of mechanisms for change provides a convenient entry to a careful exposition and analysis. *Climate Actions* offers a wealth of ideas and information for anyone seeking to make actions as effective as possible."

—Brian Martin, *Emeritus Professor, University of Wollongong, Australia*

"Rooted in social movement scholarship, and based on an original new survey of climate groups, Climate Actions offers an up-to-date, comprehensive analysis. Highly recommended for anyone interested in expanding and deepening the climate movement."

—Juliet Schor, *Professor of Sociology, Boston College, USA*

Laurence L. Delina

Climate Actions

Transformative Mechanisms for Social Mobilisation

Laurence L. Delina
Frederick S. Pardee Center for the
  Study of the Longer-Range Future
Boston University
Boston, MA, USA

ISBN 978-3-319-91883-9   ISBN 978-3-319-91884-6   (eBook)
https://doi.org/10.1007/978-3-319-91884-6

Library of Congress Control Number: 2018941872

Cover illustration: Pattern adapted from an Indian cotton print produced in the 19th
century

Printed on acid-free paper

This Palgrave Pivot imprint is published by the registered company Springer International
Publishing AG part of Springer Nature
The registered company address is: Gewerbestrasse 11, 6330 Cham, Switzerland

To Matheo Rain, Michaela Marie, Sophia Lucylle, Ian Gabriel and Michelle Lorraine, you, most likely, will inherit an altered future Earth, but you should know that some of us fought hard for it to be just, fair, and sustainable as much as possible.

PREFACE

Keeping fossil fuels in the ground and accelerating just transitions to sustainable energy systems remain essential in addressing our collective climate challenge. Despite the common aspirational goals agreed in the Paris Agreement on climate change, climate change remains a challenge that must be addressed at its core, which means the rapid reduction of anthropogenic greenhouse gas (GHG) emissions. The need for rapid GHG reduction continues to underscore climate activism, which now must become even stronger. This book provides strategies for making and doing climate actions, based on insights from 2014 and 2017 international surveys of contemporary social action groups. These distilled approaches vary across respondents, groups, and places, and some activist groups have already adopted them. But these strategies still offer opportunities for stocktaking and acquiring a foothold, so that activism can be more effective and the common good can prevail. Climate actions can be designed around relationships (relating), moral and value-based messages (messaging), alternatives (visioning), diversity (webbing), and communication (interacting). This book synthesises these critical mechanisms in strengthening the diverse, plural, multi-scale, and heterogeneous, yet inextricably linked climate action movement.

The many lectures, seminars, workshops, and courses I've attended, as well as the chats and conversations on benches and at dining tables, cafes, beaches, and in hallways in many places—from Sydney to Boston to Munich to Bangkok to Manila and elsewhere—have been key in shaping my understanding of climate actions. Throughout my travels, I learned from many fine people. I will attempt to mention them all here, but I hope that those I may neglect to acknowledge can forgive my poor memory.

This book had its beginning in Sydney, Australia, where I wrote my Ph.D. thesis. This book is the second part of that work and a response to a basket of climate mitigation strategies produced from a *Gedankenexperiment* of what can be learned from rapid wartime mobilisation. That work has been published as a book in 2016. For the second part of my Ph.D. thesis, I worked on the research question: how to produce democratic climate actions. I'm grateful to my Ph.D. supervisor and mentor Mark Diesendorf, who was the first to suggest I look at climate activism. That research focused at two opportunities—one on lessons to be learned from select moments of large-scale social movements (published in *Carbon Management*—see Delina, L, Diesendorf, M & Merson, J, 2014, vol. 5, pp. 397–409), and another on lessons to be learned from contemporary social action groups (published in *Interface: A Journal for and about Social Movements*—see Delina, L & Diesendorf, M, 2016, vol. 8, pp. 117–141). The International Center on Nonviolent Conflict (ICNC) supported this work through a predoctoral fellowship award. I thank ICNC for its generosity, particularly, Maciej Bartkowski for his comments on my writing.

This book expands on my *Interface* article and includes updates in the rapidly changing landscape of the climate action movement with insights drawn from a new survey data. This book benefited so much from these surveys. These data would not be available if not for the time of my generous respondents from around the world. The insights they shared form the core of this work and this book could not have been written without them.

From Sydney, I took this work to Harvard University, particularly at the Program on Science, Technology and Society at the Kennedy School, where I refined my arguments as a Visiting Fellow in the Spring of 2013 and Spring of 2016. I was very fortunate to learn from Sheila Jasanoff during these visits. Subsequently, a Carson Fellowship at the Rachel Carson Center at Ludwig Maximilian University in Munich gave me uninterrupted time to complete this research. I am grateful for the opportunity to be in such a vibrant intellectual community as I finished this work and also completed another book, *Accelerating Sustainable Energy Transitions*. I thank Christof Mauch, Helmuth Trishler, Arielle Helmick, and Carmen Dines for making it happen. My academic home since 2015, the Frederick S. Pardee Center for the Study of the Longer-Range Future at Boston University, has always been supportive of my work. I thank my postdoc supervisor, Center Director and Prof. Anthony Janetos, for his unwavering support. I'm also particularly grateful to Cynthia Barakatt, who for the nth time, read, edited, and commented on my work. I also thank my commissioning editor at Palgrave Macmillan, Rachael Ballard, and two anonymous reviewers for their helpful comments.

My friends and colleagues also deserve my gratitude: Franziska Mey, Vipra Kumar, Alicia Bergonia, Nahid Sultana, Long Seng To, John Connors, Matthew Burke, Jennie Stephens, Peachie Ann Aquino, George Manzano, Sheila Siar, Angeline Rodriguez-Burling, Bobby Wengronowitz, Joanna Bahian, Rima Alfafara, Ever Simonsson, Roditt Cruz-Delfino, Rufa Cagoco-Guiam, Remedios Pineda, Ma. Rudette Dardo-Palomar, Eden Joy Alperto, Beverly Valdez, Merlyn Jarrell, Ricardo Samanion, Jose Tenecio Jr., Mary Ann Frugalidad-Latumbo, Rhea Venus-Dela Cruz, Alma Dolot, Allan Lao, Aynee Triunfante, and Charmae Kacir. I also thank my family; it is them I cherish the most. I dedicate this work to my nephews—Matheo Rain and Ian Gabriel—and nieces—Michaela Marie, Sophia Lucylle, and Michelle Lorraine.

San Isidro, Sto. Niño,   Laurence L. Delina
South Cotabato, Philippines
28 March 2018

CONTENTS

LIST OF TABLES

CHAPTER 1

Introduction

Abstract While the Paris Agreement can be hailed as an important milestone for global climate action, addressing the climate challenge by curtailing future emissions remains an important agenda for climate activism. More than before, the climate action movement needs to be strengthened of its campaigns, tactics, and strategies. Despite the heterogeneity of the many actors comprising this movement and the variations in their campaigns, tactics, and strategies, each of these actors can be strengthened in its capacity for a series of and parallel climate actions, in terms of better understanding of climate-related issues, logical response particularly its solutions, and engagement through behavioural changes in consumption and supporting alternatives to fossil-based energy regimes.

Keywords Climate actions · Paris Agreement · Social movement Climate activism

The 2015 Paris Agreement on climate change—the result of a long journey that begun in Rio de Janeiro in 1992 with significant stops in Kyoto, Bali, Copenhagen, and elsewhere—ideally should have included a 'legal instrument...to achieve...stabilization of greenhouse gas concentrations in the atmosphere that would prevent dangerous anthropogenic interference with the climate system...[which] should be achieved within a time frame sufficient to allow ecosystems to adapt naturally to climate

© The Author(s) 2019
L. L. Delina, *Climate Actions*,
https://doi.org/10.1007/978-3-319-91884-6_1

1

change, to ensure that food production is not threatened and to enable economic development to proceed in a sustainable manner' (Article 2 of the United Nations Framework Convention on Climate Change (UNFCCC) (United Nations 1992). Despite significant advances in scientific understanding that inadequate climate change mitigation could increase 'the likelihood of severe, pervasive and irreversible impacts for people and ecosystems' (IPCC 2014: 8), the international ambition to mitigate climate change remains elusive.

The Paris Agreement is a vital step to meet this ambition, but it still misses on key areas of climate actions required to stave off the worst impacts. The Nationally Determined Contributions (NDCs) incorporated into the Agreement, even if attained, are still too weak to ensure that dangerous climate change is avoided (Climate Action Tracker 2015). Also missing is the strong review and enforcement mechanism to ensure the Parties are locked into meeting Article 2 of the UNFCCC and the 2015 aspirational, but not mandatory, contributions: 'Holding the increase in the global average temperature to well below 2 °C above pre-industrial levels and to pursue efforts to limit the temperature increase to 1.5 °C above pre-industrial levels.' Absent these, and just like its Kyoto Protocol predecessor, the Paris Agreement will fail to deliver the necessary climate actions and, without strong and active public engagement, it will also fail to pressure decision-makers in regional, national, and subnational jurisdictions.

Given how high the stakes are, strengthened climate actions remain vital in disrupting the current order. Peoples must be mobilised so they move from the balcony to the barricades, bystanders are transformed into activists, and adversaries are demobilised. Climate actions remain necessary until fossil fuels are kept in the ground, organised resistance from incumbent industrial actors and their allies in politics and the media is dismantled, vulnerable countries are provided necessary support to adapt to climate change and contribute to its mitigation, and a just transition to a sustainable economy—one that is not hinged on industrial capitalism, unfettered accumulation and growth, and the monetised terms of economic value—is addressed at a pace suggested by our improved understanding of climate science.

When governments fail to effectively mitigate climate change, the climate action movement must strengthen the pressure it places on power-holders. The climate action movement—comprise of diverse and heterogeneous peoples, climate action groups, networks, alliances, and coalitions (cf. Giugni and Grasso 2015)—seems to be the only hope to achieve this change. While this movement may be considered transnational because of the boundary defiance of the climate challenge,[1] it is still (and must be) grounded in the locality of individual's and small groups' social actions. Local actions are not only compatible with transnational climate action; they are also intertwined. Nonviolent strategies, tactics, and other campaign approaches must be strengthened and aligned to meet these unachieved goals.[2]

Climate activism has already driven climate actions. It was essential, among others, to bring the climate issue into popular understanding and mobilise peoples into climate actions. Examples of this activism include the 2014 People's Climate March,[3] stopping the Keystone Pipeline project, applying pressure to universities, churches, and other institutions to divest from fossil fuels, and propelling the ongoing small-scale, local-based transitions through households, communities, and neighbourhoods, local governments and cities, and businesses. This cache of successes demonstrates that the social movement for climate actions has achieved change, albeit slowly. Indeed, the fossil fuel regime—industries and their supporters in governments and media—still occupy prominence in decision-making.

Studies about the climate action movement have already gained salience. The literature on its strategies (e.g. Delina et al. 2014; Klein 2014; Diesendorf 2009), ethnography (e.g. Rosewarne et al. 2014; Foran 2014),

---

[1] Transnational climate activists (or climate action proponents), following Sidney Tarrow's (2005) conceptualisation, refer to 'people and groups who are rooted in specific national contexts, but who engage in contentious political activities that involve them in transnational networks and contacts.'

[2] Sceptics of nonviolent social movements may argue that violent insurgencies could provide immediate results and, therefore, are best suited for effective climate actions. Empirical data, however, reveals that nonviolent civil resistance is more successful than violent resistance in terms of historical track records, the degree of popular participation, and the lower levels of suffering (see Gleditsch and Celestino 2013; Chenoweth and Stephan 2011).

[3] On 21 September 2014, the power of the networked approach was evidenced in one of the largest gatherings of the climate action movement where 350.org and a number of other organisations coordinated the so-called People's Climate March. *The Guardian* reported that the campaign involved an estimated 570,000 people taking part in 2700 simultaneous events in 161 countries. In its culminating activity in the streets of New York, an estimated 400,000 people and 1573 groups were represented, according to the organisers.

and solidarities and networks (e.g. Routledge 2012; Featherstone 2005, 2008; Cumbers et al. 2008), among others, has been enriched. These works have, in effect, also expanded the already established corpus of knowledge about social movements (e.g. Tilly 1995, 2002, 2008; Moyer 1987, 2001; Mann 1993; Ganz 2004), of which climate actions are part of. This book contributes to this corpus by highlighting the continuing relevance of climate actions—including the mobilisations of peoples and groups—to drive behavioral, sociotechnical, political, and economic changes.

Climate mitigation, indeed, has to be understood not only as a technological transition project but also inclusively, if not more importantly, as a work on social-political-economic as well as behavioral transformations. Acknowledging the sociotechnicality of climate actions, this book highlights the insights drawn from surveys of contemporary social change groups—whom I solicited for their experiential knowledge in 2014 and 2017–2018—as key complements to an already much richer literature on the technologies of the transition. Although admittedly climate action groups adopt different approaches and pathways to the transition to a low carbon future, their heterogeneous insights could generate new ways of thinking about making and producing effective climate actions. This book confirms some of the fundamental approaches in creating effective social mobilisation campaigns effective and, at the same time, underscores their continuing utility for spurring future climate actions.

Climate actions can be articulated as the campaigns, tactics, and strategies to achieve the active participation and engagement of publics, small groups, and communities primarily for, but not strictly limited to, collective climate mitigation. At the centre of this thinking are the publics, engaging in transformative climate actions. These actions need to go well beyond the traditional approaches of lobbying power-holders and voting in elections, and, instead, follow and extend what Gene Sharp (1973b) has conceptualised as citizen nonviolent actions. According to Sharp, these actions are classified into: (1) Protest and persuasion, non-cooperation, and intervention (e.g. active participation in mass rallies, teach-ins, media campaigns, strikes, boycotts, and street demonstrations); (2) Other forms of social, political, and economic non-cooperation (e.g. fundraising, communicating messages, and refusing to cooperate with governments); and (3) Community demonstrations of alternatives.

Peter North (2011) has developed a taxonomy to relate the concept of social mobilisations with climate actions. According to North, climate activists engage the public using two broad approaches: outward-focused and prefigurative. On one hand, 'outward-focused' climate activism is seen through confrontational protests, demonstrations, and lobbying marches to put pressure on political elites to act. This resonates well with the first two of Sharp's (1973b) classification mentioned above. On the other hand, 'prefigurative activism' engages beyond the 'what needs to be done' question and instead elaborates action that goes beyond diagnosis and prognosis (cf. Snow 2013). This focuses on the individual and community levels to develop community-based solutions and opts for smaller scale, grass roots-oriented technologies such as rooftop solar panels. This mirrors Sharp's (1973b) third group of activists. Those who direct their policy reform practices towards community-level transitions also exemplify this group. Their approach resonates with the concept of new technology development at niches often found in local communities through the provision of 'spaces for innovation' in developing low carbon lifestyles that might later diffuse to the mainstream (cf. Seyfang et al. 2013; Foxon 2013; Hopkins 2008). Yet, these community approaches involve not only technological shifts; they also require social relationship transformations, stronger institutions, and influence over local power balances (Cinderby et al. 2015) inasmuch as they also must led into behavioural changes, particularly with regard to consumption habits.

Using these concepts, aided by the social movement literature (e.g. Lorenzoni et al. 2007; Moser 2009, 2007a, b, c; Moser and Dilling 2007; National Research Council 2002), climate actions can be described as a series of or parallel activities that enable publics to: (1) Engage with and gain understanding of the climate change issue, especially its drivers, impacts, and solutions; (2) Engage with a logical response to the climate issue, including concern over its effects or interest towards its solutions; and (3) Involve themselves actively in climate actions, changing their consumption behaviours, and adopting, supporting, or demonstrating the many alternatives to existing fossil-based energy regime.

The first two of these activities have already been accomplished as evidenced by the large majority of the public who support climate actions (e.g. the huge turnout during the 2014 People's Climate March). However, a strategy has yet to be widely and effectively deployed for achieving the third condition—that is, bringing to scale community energy transitions that have already been demonstrated in many locations. Fortunately for the climate action movement, several practical examples of such approaches exist, including the German *Energiewende*[4] and the Danish energy transitions. Backed by community renewable energy, these large-scale climate actions were duly accomplished. One of the principal directions this book is attempting to track, thus, is *how* the processes of social mobilisations can be ramped up so that opportunities for similar wide-ranging climate actions can be suggested to the public.

Climate actions can be delineated according to campaigns, tactics, and strategies. While they may sound similar, they vary conceptually. 'Campaigns' could refer to a series of observable continual tactics in pursuit of a political objective (cf. Ackerman and Kruegler 1994). 'Tactics' could refer to individual steps or tools used in carrying out a strategy (Diesendorf 2009: 233). These are 'specific activities with which you implement your strategy—targeted in specific ways and carried out at specific times' (Ganz 2006a). 'Strategy' may refer to the planning and conduct of long-term campaigns to achieve broad goals (Diesendorf 2009: 234). Marshal Ganz (2006a) suggests a strategy to be about *how* to turn 'what you have' into 'what you need' to achieve 'what you want.' A strategy, Ganz (2006a) expands, is the 'conceptual link leaders make between the places, the times and the ways they mobilize and deploy resources and the goals they hope to achieve by this mobilization.' A strategy is 'both analytic and imaginative, figuring out how we can use our resources to achieve our goals' (Ganz 2006b). For the contemporary climate action movement to be strengthened in its campaigns, tactics, and strategies, it must seek opportunities that would ensure the effectiveness of its climate actions.

Measuring the effectiveness of any social action campaign, tactic, and strategy, however, is a controversial exercise. In climate activism (and in other social actions), where actions are best seen as nodes in a continuum that take place over long periods, effectiveness is hard to measure (cf. Tilly 2008). Since some of these actions may fail and some of the

---

[4] However, *Energiewende* also has its shortcomings. Haas and Sander (2016), for instance, suggest—given that the share of renewable energies in total energy consumption is only around 12% and that there have been weak advances in other energy services particularly in heating and transport—it is more than an electricity transition rather than a broader energy transitions. Nonetheless, *Energiewende* offers an evidence that communities can be key actors in large-scale transformations.

goals may change, it is not unusual that some goals along this continuum are unmet (Rosewarne et al. 2014: 81–84). The effectiveness of climate actions, however, can be measured, following one suggestion—that by Charles Tilly (2002, 2008), using 'mechanisms' that are visible at the interactional level. This can be achieved by looking at the dynamics of contentious gatherings and public performances—the campaigns, the tactics, and the strategies (cf. Collins 2010).

According to Tilly (2002, 2008), the mechanisms surrounding any social action remain spatially and temporally constant (cf. Collins 2010). A social action is about communicating what Tilly (2002, 2008) calls WUNC, an acronym for worthiness, unity, numbers, and commitment. Worthiness is displayed in the campaigners' sober demeanors by showing the public that they are a decorous people. Unity is achieved by, for example, marching in ranks, singing, and chanting together. Numbers are shown in headcounts, signatures on petitions, and/or the capacity to fill streets. Commitment is portrayed in campaigners' resolve and willingness to undergo hardships, such as braving bad weather or defying state repression. WUNC, which clearly underlines the principle of nonviolence at all times, establishes the appeal of social actions and legitimates social action movements. As a result, the movements may grow and become effective modes of modern politics.

To ascertain the effectiveness of climate actions, I have followed Tilly's focus on 'mechanisms' in my work. In one of my papers (i.e. Delina et al. 2014), my co-authors and I refer to these 'mechanisms' as dominant dynamics, patterns, elements, and key tensions in a social action campaign, tactic, or strategy. We have identified these mechanisms using comparative analysis of episodes of four historical social action campaigns: the 1930 events in the Indian Freedom Struggle; the 1955–1956 bus boycott catalysing the modern African-American Civil Rights Movement; the anti-Marcos rallies culminating in the 1986 Philippine People Power Revolution; and the 1988–1990 campaigns known as the Burmese Uprising. In this paper, I suggest that the mechanisms for effective (and ineffective) campaigns have revolved on: building a new collective identity and a unified regime alternative, communicating the moral message, and enrolling a diversity of participants and networking. Following this, the indicators for 'effective social action,' thus, can be based on their outcomes. These outcomes, however, are spatially and temporally dependent; but, they can be broadly described as those moments when the target audience: (1) funds and/or provides other resource support; (2) joins the social action group themselves; and/or (3) actively engages with group activities. The final litmus test for effectiveness, of course, is when the stated objectives of the social action are duly realised.

To triangulate my descriptions of effective social actions, I have conducted two online qualitative surveys of contemporary social action groups. These surveys were administered over two weeks in November 2014 and over a month in December 2017 to January 2018. In both instances, the surveys use the same instrument to determine how contemporary social action groups build, implement, monitor, and reflect upon their campaigns, tactics, and strategies; that is, how they 'measure' their effectiveness. It is key to note that these surveys were not probability studies (that require random samples to statistically represent a given population) but a purposive study (where participants are recruited based on a specific purpose and with a specific target audience).

In these surveys,[5] contemporary social action groups refer broadly to groups whose activities involve campaigns, tactics, and strategies to affect social change. They include, among others, labour unions, professional groups, faith-based organisations, women's groups, and environmental groups. Their responses to the surveys were interpreted as verbal, rather than numerical statements. These responses are reported as unabridged verbatim quotes as much as possible to preserve as much authenticity of the statements and narratives and to emphasise contextual respondent-level commentaries. Although the respondents were let to 'speak for themselves' to reduce interpretive bias, their texts had also been analysed and critiqued whenever appropriate. In this book, I refer to respondents' quotes using the following crude shorthand: SAE for a social action group whose focus of campaign is on the environment, SAJ for a social justice group, and SAO for 'other' groups.[6] The respondents identified themselves as belonging to one of these categories. Based on the survey data, I have identified five key mechanisms for making effective climate actions: relating, messaging, visioning, webbing, and interacting.

The first of these identified mechanisms, called *relating*, is about enhancing face-to-face meetings to turn talk into effective climate actions.

---

[5] Appendix A shows the survey questionnaire; Appendix B provides notes on the research instrument, and statistical treatment and reporting; Appendix C describes the respondents.

[6] See Table C.1 for the descriptions of these groups.

While this opportunity is already present among climate action groups, there are still gaps that must be understood, filled, and reflected upon. This chapter asks how climate action campaigners can orient their conversations to achieve stronger engagement, what are the strengths and limits observed by contemporary social action groups when building relationships with their publics, how can these strengths be translated into other spaces and audiences, what do the weaknesses tell us, and what are the emergent practical lessons from these experiences.

The second mechanism, *messaging*, is about recognising that psychological stimuli can change people's attitudes towards social change; yet their use can also open tensions that can backfire when not managed well. The strategic use of these stimuli in social mobilisation often depends upon how skillfully campaigners could orient their messages vis-à-vis the steadfast values and moral concerns of their audience. While evoking morality in campaigns can be effective in mobilisation, it, too, may backfire. This chapter asks how future messages can be adapted to the audience's moral compass, what are the pitfalls of morally charged messages for climate actions, and how could activists—using examples from other mobilisations—address these.

In the continuum describing social movements, the most important phase is when a large majority of people engage in social actions including changing their personal behaviours and mindsets. This chapter on *visioning* acknowledges that people could be provided with this new sense of collective identity and ownership when campaigners provide them a clear and unified regime alternative. Here, I ask how climate action campaigners can construct a new vision for a stable climate era; what are the complexities involved in these processes; what examples are blossoming elsewhere; how do they materialise and how could these be used as visioning materials for climate actions; and what are the weaknesses of this strategy as experienced in other social actions.

Fostering diversity of participation is vital, especially in climate actions where multi-scale and multilevel approaches are imperative. Diversity of participation is achieved when the campaign involves as many participants as possible and encompasses genders, socio-economic classes, religions, beliefs, etc. Equally essential for success, however, is linking these heterogeneous, multi-scale actions. To that end, this chapter calls for *webbing* and, so, asks: how can networks be established in ways that encourage, engage, and empower campaigners and groups instead of weakening and dispiriting them? It describes how non-hierarchical and centreless forms of climate actions can be triggered, sustained, and scaled as a collective. The chapter further asks: as these dynamics play out, what can contemporary social action campaigns tell us about group independence and dependence, and what does this mean when webbing divergent, multi-scale, and transnational, yet grass roots, climate actions?

The final chapter deals with *interacting*, a shorthand for reaching out to a large proportion of the population. Numbers remain imperative for effective mobilisations; thus, anonymous interactions among peoples of like minds and other commonalities remain an imperative for climate actions. Extensive media coverage and critical social problem explanations in popular media had helped in speeding large-scale mobilisations but using traditional media may not work for soliciting effective climate actions. Indeed, low media attention and unfriendly media stance over climate issues have remained prominent. In that regard, this chapter asks: how contemporary action groups achieve effective public communication in cases of lack of coverage and inaccurate coverage in the media; how can these communication strategies be translated into climate actions; what are the strengths and limitations of social media in mobilisation; and how do contemporary groups use them effectively.

REFERENCES

Ackerman, P., & Kruegler, C. (1994). *Strategic Nonviolent Conflict: The Dynamics of People Power in the Twentieth Century*. Westport, CT: Praeger.

Chenoweth, E., & Stephan, M. J. (2011). *Why Civil Resistance Works: The Strategic Logic of Nonviolent Conflict*. New York: Columbia University Press.

Cinderby, S., Haq, G., Cambridge, H., & Lock, K. (2015). Building community resilience: Can everyone enjoy a good life? *Local Environment, 21*, 1252–1270.

Climate Action Tracker. (2015, November 13). *G20—All INDCs in, but Large Gap Remains*. Berlin, Germany: Climate Analytics and Ecofys; Potsdam, Germany: Potsdam Institute for Climate Impact Research; and Cologne, Germany: NewClimate Institute.

Collins, R. (2010). The contentious social interactionism of Charles Tilly. *Social Psychology Quarterly, 73*, 5–10.

Cumbers, A., Routledge, P., & Nativel, C. (2008). The entangled geographies of global justice networks. *Progress in Human Geography, 32*, 183–201.

Delina, L., Diesendorf, M., & Merson, J. (2014). Strengthening the climate action movement strategies from histories. *Carbon Management, 5*, 397–409.

Diesendorf, M. (2009). *Climate Action: A Campaign Manual for Greenhouse Solutions*. Sydney, Australia: UNSW Press.

Featherstone, D. (2005). Towards the relational construction of militant particularisms: On why the geographies of past struggles matter for resistance to neoliberal globalisation. *Antipode, 37,* 250–271.

Featherstone, D. (2008). *Resistance, Space and Political Identities: The Making of Counter-Global Networks.* Oxford, UK: Wiley-Blackwell.

Foran, J. (2014). "¡Volveremos!/we will return": The state of play for the global climate justice movement. *Interface, 6,* 454–477.

Foxon, T. J. (2013). Transition pathways for a UK low carbon electricity future. *Energy Policy, 52,* 10–24.

Ganz, M. (2004). Why David sometimes wins: Strategic capacity in social movements. In D. M. Messick & R. M. Kramer (Eds.), *The Psychology of Leadership: New Perspectives and Research.* Malwah, NJ: Lawrence Erlbaum Associates, Publishers.

Ganz, M. (2006a). Strategy, deliberation and meetings. In *Organizing Course Notes.* Cambridge, MA: Harvard Kennedy School. http://bit.ly/1Dx2ppv.

Ganz, M. (2006b). Mobilizing power: Analysis, strategy, deliberation. In *Organizing Course Notes.* Cambridge, MA: Harvard Kennedy School. http://bit.ly/1xcbrGW.

Giugni, M., & Grasso, M. T. (2015). Environmental movements in advanced industrial democracies: Heterogeneity, transformation and institutionalization. *Annual Review of Environment and Resources, 40,* 337–361.

Gleditsch, K. S., & Celestino, M. R. (2013). Fresh carnations or all thorn, no rose? Non-violent campaigns and transitions in autocracies. *Journal of Peace Research, 50,* 385–400.

Haas, T., & Sander, H. (2016). Shortcomings and perspectives of the German Energiewende. *Socialism and Democracy, 30,* 121–143.

Hopkins, R. (2008). *The Transition Handbook: From Oil Dependency to Local Resilience.* Totnes, Devon, UK: Green Books.

Intergovernmental Panel on Climate Change (IPCC). (2014). *Climate Change 2014: Synthesis Report of the Fifth Assessment Report of the IPCC.* The Core Writing Team, R. K. Pachauri & L. Meyer (Eds.). Switzerland: IPCC.

Klein, N. (2014). *This Changes Everything: Capitalism vs. the Climate.* New York: Simon & Schuster.

Lorenzoni, I., Nicholson-Cole, S., & Whitmarsh, L. (2007). Barriers perceived to engaging with climate change among the UK public and their policy implications. *Global Environmental Change, 17,* 445–459.

Mann, M. (1993). *The Sources of Social Power, Volume 2: The Rise of Classes and Nation-States, 1760–1914.* Cambridge, UK: Cambridge University Press.

Moser, S. C. (2007a). Communication strategies to mobilize the climate movement. In J. Isham & S. Waage (Eds.), *Ignition: What You Can Do to Fight Global Warming and Spark a Movement* (pp. 73–95). Washington, DC: Island Press.

Moser, S. C. (2007b). In the long shadows of inaction: The quiet building of a climate protection movement in the United States. *Global Environmental Politics 7*, 124–144.

Moser, S. C. (2007c). More bad news: The risk of neglecting emotional responses to climate change information. In S. C. Moser & L. Dilling (Eds.), *Creating a Climate for Change: Communicating Climate Change and Facilitating Social Change* (pp. 64–80). Cambridge, UK: Cambridge University Press.

Moser, S. C. (2009). Costly knowledge—Unaffordable denial: The politics of public understanding and engagement on climate change. In M. T. Boykoff (Ed.), *The Politics of Climate Change: A Survey* (pp. 155–181). Oxford, UK: Routledge.

Moser, S. C., & Dilling, L. (Eds.). (2007). *Creating a climate for change: Communicating climate change and facilitating social change*. Cambridge, UK: Cambridge University Press.

Moyer, B. (1987). *The Movement Action Plan: A Strategic Framework Describing the Eight Stages of Successful Social Movements*. The Social Movement Empowerment Project. http://bit.ly/1f9KI6p.

Moyer, B., McAllister, J., Finley, M. L., & Soifer, S. (2001). *Doing Democracy: The MAP Model for Organizing Social Movements*. Gabriola Island, BC: New Society Publishers.

National Research Council. (2002). *New Tools for Environmental Protection: Education, Information, and Voluntary Measures*. Washington, DC: National Academy Press.

North, P. (2011). The politics of climate activism in the UK: A social movement analysis. *Environment and Planning A, 43*, 1581–1598.

Rosewarne, S., Goodman, J., & Pearse, R. (2014). *Climate Action Upsurge: The Ethnography of Climate Movement Politics*. Abingdon, Oxfordshire, UK and New York: Routledge.

Routledge, P. (2012). Translocal climate justice solidarities. In J. S. Dryzek, R. B. Norgaard, & D. Schlosberg (Eds.), *The Oxford Handbook of Climate Change and Society*. Oxford, UK: Oxford University Press.

Seyfang, G., Park, J. J., & Smith, A. (2013). A thousand flowers blooming? An Examination of Community Energy in the UK. *Energy Policy, 61*, 977–989.

Sharp, G. (1973a). *The Politics of Nonviolent Action: Part One, Power and Struggle*. Boston, MA: Porter Sargent.

Sharp, G. (1973b). *The Politics of Nonviolent Action: Part Two, the Method of Nonviolent Action*. Boston, MA: Porter Sargent.

Snow, D. A. (2013). Framing and social movements. In D. A. Snow, D. della Porta, B. Klandermans, & D. McAdam (Eds.), *The Wiley-Blackwell Encyclopedia of Social and Political Movements*. Malden, MA: Blackwell.

Tarrow, S. (2005). *The New Transnational Activism*. Cambridge, UK: Cambridge University Press.

Tilly, C. (1995). *Popular Contention in Great Britain, 1754–1834.* Cambridge, MA: Harvard University Press.

Tilly, C. (2002). *Stories, Identities and Political Change*. New York: Rowman & Littlefield.

Tilly, C. (2008). *Contentious Performances*. Cambridge, UK: Cambridge University Press.

United Nations. (1992). *United Nations Framework Convention on Climate Change* (FCCC/INFORMAL/84). http://unfccc.int/resource/docs/convkp/conveng.pdf.

CHAPTER 2

Relating

Abstract Face-to-face interactions remain key in turning talk into effective climate actions. While this opportunity is already present among climate action groups, there are still gaps that need to be understood, filled, and reflected upon when messengers relate with their audiences. This chapter asks how can climate action campaigners orient their dialogues and conversations to ensure stronger engagement with their audience, what are the strengths and limits observed by contemporary social action groups when building relationships with their publics, how can these strengths be translated into other spaces and audiences, and what the weaknesses tell and what practical lessons emerged from these experiences.

Keywords Climate communication · Face-to-face activism
Public engagement

I say that: "The air I'm breathing in this room is chemically different than it was when I was born" (i.e. 400 ppm today, 360 ppm at my birth). I stress that we have "changed the basic chemistry of our planet". And contextualize it by taking a moment to talk about "this tiny rock hurtling through some random corner of space" and "its thin biosphere that supports all life as we know it," etc. I really speak slowly - with intentional pauses - letting the gravity of our situation sink in. I find it's useful to spend 2-3 min on "the set up"... to give people a chance to step back for

© The Author(s) 2019   15
L. L. Delina, *Climate Actions*,
https://doi.org/10.1007/978-3-319-91884-6_2

a second and really consider where we're at. I've been noticing that people then become a lot more receptive to my pitch.
—Respondent from a Canada-based group formed in 2013 that lobbies local governments to pass laws that would require gasoline retailers to place climate change and air pollution labels on their gas pump nozzles.

In the winter of 1978, the Organisation for Information about Nuclear Power in Denmark (OOA) mobilised grass-roots support to push the Danish government to drop a nuclear power proposal. Volunteer campaigners distributed 'Denmark without nuclear power' leaflets and visited a majority of Danish homes to explain why nuclear power is not the future for the Danes. In these visits, which continued until 1980, OOA emphasised broad, popular information on energy and environmental questions surrounding nuclear energy (Læssoe 2007). These home visits became OOA's primary and most important strategy that eventually led to the mobilisation of substantial and strong public support against the nuclear option in the country. As a result of these face-to-face campaigns, the Danish Government was forced to drop the nuclear option (Læssoe 2007).

The success of face-to-face engagement by missionary religions provides another example. Erik Assadourian (2013) from the Worldwatch Institute notes how face-to-face interactions have been helping many missionaries to effectively preach across different eras, values, cultural identities, and geographies. Door-to-door preaching as practised by the Mormons and Jehovah's Witnesses, Assadourian (2013) cites as examples, proved effective in increasing their membership base. Their approaches are built solidly on structured conversation guides allowing an environment for question-and-answer (Q&A) to thrive. Many other Christian reformers, such as the Salvation Army and the Knights of Columbus, have also built global reach using these techniques, aside from creating social programmes that follow the communitarian model such as food and shelter provision (Assadourian 2013).

Face-to-face conversations with communities, including engagement with local politics and activities, indeed, are vital mobilisation approaches according to survey respondents. Most respondents agree that education, information, and awareness campaigns remain most effective in soliciting support ($N=47$, median$=4$, Interquartile range

(IQR) = 1^1). SAE26[2] puts it this way: "Instead of getting folks to come to you, we have found it somewhat effective for us to go to them..." These approaches include lectures, seminars, trainings, workshops, and pamphlet distribution, among others. Twenty-eight respondents or 60% rate this approach as 'effective' while 18 or 38% see them as 'very effective.'

Face-to-face engagement is considered more effective for the following reasons. First, it is more personal and therefore allows the audience to deeply understand the issues. Second, it helps campaigners identify the best ways to frame an issue. Within the context of social movements, 'framing refers to meaning construction engaged in by movement adherents relevant to the interests of movements and the challenges they mount in pursuit of those interests' (Snow 2013). Framing, through images, personal stories and narratives, tone of voice, words, and other signals, provides essential contexts for people to make sense of the issue—and is effective only with face-to-face interactions. This personalised delivery triggers a cascade of responses and can prime a particular audience for effective action (Moser and Dilling 2007; Fisher and McInerney 2012). Third, non-verbal cues allow campaigners to gauge how the information is being received in real time and to respond accordingly. The campaigner, therefore, has ample means to become creative while learning from the ongoing interaction. Fourth, direct engagement allows for a dialogue to emerge since these free flowing, yet structured, interactions facilitate Q&A. A two-way exchange also strengthens the trust between individuals and can go a long way towards engaging more audiences.

Alignment is key for effective face-to-face mobilisations. Campaigners must link their interests and goals with those of their audiences, so they too can contribute to climate actions. The dynamics of alignment involve processes of 'bridging' (Snow et al. 1986), which refers to linking two or more ideologically congruent but structurally disconnected frames regarding a particular issue. To achieve 'bridging,' the extent to which climate actions are framed must be experientially commensurate with the audience's past and present lives. In talking with people, SAE14 emphasises the importance of "making [the conversations] very personal"

---

[1] See Appendix B for notes on statistical treatment and reporting.

[2] See Table C.1 for the description of the social action groups.

and that "knowing your audience remains paramount." SAE37 advises: "What matters is good research, good information, [and] the ability to relate the information to people's experience. If they don't see a connection, they have no reason to be interested." In these interactions, SAE04 suggests the campaigner to "point out how an issue affects each person: their health, their wallet, their children, the things they value and enjoy since 'everyone is an environmentalist when the issue affects them.'" SAE10 quips: "We have to somehow appeal to people on the basis of what directly impacts them. So to find that button is the key." In short, conversation ideas must be personalised "to get people on board" (SAE19) and can be done, for example, by using personalised probing questions such as "What if this happened to you? Would you know what to do?" (SAE19).

The examples one sets also can wield a lot of influence. The psychology of tribes suggests that messages resonate first within established groups rather than at the bigger movement per se (cf. Saunders 2008). A social action becomes effective, according to majority of survey respondents, if people see a member of their family or relatives involved in these actions ($N=40$, median$=4$, IQR$=1$). Eighteen respondents or 45% state their campaigns have been 'effective' by mobilising their own family members first, while 14 or 35% declare this directed approach as 'very effective.' SAE35 notes: "[Working] within existing relationships and networks – these are the area where individuals have the greatest sphere of influence…the attention of close friends and family to particular social and political causes is a powerful motivator to support those causes."

In the past, ties with friends and family members had affected participation in social actions as demonstrated, for example, in the cases of the Mississippi Freedom Summer (Fernandez and MacAdam 1988), the East German Revolution of 1989 (Opp and Gern 1993), and the Dutch Peace Movement (Klandermans and Oegema 1987). Most survey respondents strongly agree that their approaches were 'effective' if people see their friends involved in their groups' campaigns ($N=44$, median$=4$, IQR$=1$). Twenty-five respondents or 57% state campaigning within a circle of friends as 'effective,' while 17 or 39% state it has been 'very effective.' It is key to note, nonetheless, that evidence also exists suggesting weaker participant retention in cases where mobilisations had occurred within these personal connections (Fisher and McInerney 2012). The same study, instead, suggests that self-starters—those who joined through their own volition—tend to stay longer. This conclusion highlights the need to also focus mobilisation efforts to strangers, starting with nearby communities.

Most respondents strongly agree that their campaign approaches are 'effective' if campaigns are directed at the community or area where campaigners reside ($N=44$, median$=4$, IQR$=1$). Eleven respondents or 25% declare that campaigning within their own local areas as 'very effective' approach. SAO05 agrees to focus on community-based campaigning "to build resilience and increase awareness and action." SAO03 and SAE08 note these campaigns could include "kitchen conversations among friends and neighbours." It could also occur at places of worship as SAO07 reports: "Church and community mobilisation is our most effective way of helping churches serve their communities and help communities see what resources they have, find hope, and develop a vision for change." According to SAO04, mobilisations can occur even at mundane everyday places "such as showing up to school board meetings... and PTA's (Parents-Teachers Associations)."

What is clear from these remarks is the imperative of identifying some sense of commonality across the audiences. Twenty-four respondents or 65% state that understanding these commonalities has been an 'effective' approach to mobilisation, while 8 or 22% even declare this to be 'very effective.' Targeting campaigns on places where social groups already exist has been, to most respondents, an effective approach ($N=42$, median$=4$, IQR$=1$). Twenty-three respondents or 55% state that working within these organised groups is an 'effective' approach, while 14 or 33% affirm this focused approach as 'very effective.'

These commonalities can be found in faith groups, cultural groups, professional associations, youth groups, and clubs. Most respondents indicate that their campaign approaches were 'effective' if they have campaigners coming from the same culture as their target audience ($N=29$, median$=4$, IQR$=0$). Twenty respondents or 69% state that this approach has been 'effective,' while 6 or 21% state that having campaigners of similar ethnicity as their target audience has been 'very effective.' Most respondents also agree that their campaigns were effective if people can see their group member or campaigner as someone who shares similar faith or religious orientation with them ($N=35$, median$=4$, IQR$=2$). SAO07 notes: "Working through churches works well because [they are] already communities that meet often and share values." 10 or 40%

and 8 or 32% of the respondents state that this approach has been 'effective' and 'very effective,' respectively.

Seeing campaigners as the same age brackets as the audience has, according to most respondents, led to 'effective' gathering of support and engaging with people ($N=36$, median$=4$, IQR$=0$). Twenty respondents or 56% state that having campaigners who are the same age as their target audience has been 'effective,' while 8 or 22% state this has been 'very effective.' Another anchor position that contemporary social action groups used to gather support and engage a wider audience is to bring campaigners to an audience where people can see the campaigners as someone belonging to their own profession. Most respondents indicate strong agreement that this approach was 'very effective' ($N=34$, median$=4$, IQR$=0$) with twenty-three respondents or 68% saying this approach has been 'effective.'

If members of the group do not necessarily share any of the above social ties, communal identities, or background characteristics, opportunities for motivating participation of disconnected individuals are also extensive (Vala and O'Brien 2007). For instance, most respondents indicate that their campaign approaches are 'effective' if people can see that members of a particular group engage in joint activities ($N=38$, median$=4$, IQR$=0$). Twenty-six respondents or 68% state that engagement in joint activities has been effective, while 8 or 21% state this has been a 'very effective' approach. Other frames used to attract disconnected individuals to the larger movement include moral shocks such as corruption or human rights violation (Jasper and Poulsen 1995) and direct appeals based on cultural or ideological alignments (Snow et al. 1986).

Face-to-face mobilisations in these tribes and communities must be about letting people learn about climate solutions particularly how to turn talk into action within their own cultural milieu. In addressing audiences, campaigners should also include discussions about the diverse, yet locally achievable climate actions and their multiple benefits, not just on the science and the impacts of climate change although, of course, these too are important. Some examples of these climate actions are demanding rapid mitigation policies from governments, supporting or joining outward-oriented climate activism, and hosting similar climate action meetings in their own households and/or local communities, schools, offices, or places of worship.

Ultimately, there is no substitute for action. Doing climate actions can unlock potential that people may not realise is possible. A clear message about what people can do, or how they can make a start is paramount. Climate actions can begin from many points since greenhouse gases are embedded in everything we use and do, but some steps are the most worthwhile, and thus make more sense. At some point, it becomes clear which steps will likely be most effective. One doable sequence that could yield the most emission reductions at the individual level is suggested in a United Nations Environment Programme (UNEP) document (Kirby 2008) and worth mentioning here. The document suggests to (1) focus on whatever takes the lion's share of one's personal emissions. Over time, these shares change and other emission areas may become more important to focus on. It is also a must to (2) avoid using or even consuming anything that will increase one's personal emissions whenever possible and to (3) choose the option that will allow one to make a reduction such as by increasing the efficiency of their activities. (4) As more efficient methods or technologies come along, individuals have to innovate in ways they contribute to climate actions.

UNEP further suggests starting with free options and working up to more expensive options as one go along. For example, if one thinks their public transport system must be replaced with less-polluting alternatives but cannot see how to afford it, then they can go for something affordable such as encouraging walking and cycling by making city streets safer for these activities. It is key, however, to note that while many climate actions are possible for individuals to have an indirect impact on reducing emissions, it is still key to do them in groups. One's small contribution may only be a drop in the ocean but many efforts taken together will definitely add and help alleviate the burden of greenhouse gas accumulation in the atmosphere. Knowing how we influence others, thus, remains crucial.

When climate actions stem from ordinary individuals and citizen groups, the results are more far-reaching. One can reach their tribes of friends and neighbours; a company to its clients; a city to its inhabitants. SAE27 notes the importance of expanding smaller actions, beyond communities: "Building a sense of efficacy from the small scale to the large scale within the grass roots and sections of the elites is vital." Solutions, however, do not come in a one-size-fits-all format. Indeed, Saunders (2008) notes that focusing on building a collective identity is not always beneficial for the larger movement in that pursuing homogeneity can be divisive. Climate actions, thus, must be tailored to individual and group circumstances, as do the types of face-to-face climate action engagement.

Many current climate action campaigns are planned, centred, and accomplished around outward-focused activism mainly through protests, demonstrations, and lobbying against fossil-based development. Among the many examples of outward-focused climate activism are the series of Keystone Pipeline protests in the USA; calls for divestment in superannuation or pension funds, colleges, and universities; direct actions against coal-fired power plants and fossil fuel extractions; and lobbying for a carbon price through legislation. This type of activism, which targets the worst, obstructionist actors for effective climate action, however, must be diversified and strengthened. This starts by identifying the obstructionist actors[3] for effective climate mitigation followed by nonviolent direct action tactics. Mark Diesendorf (2009) provides some of these tactics, which include: rallies, marches, sit-ins, and pickets; naming and shaming; shareholder actions; withdrawal of deposits; and strikes and boycotts. Other forms of outward-oriented climate activism should also be thought-out and included to increase campaign diversity.

In addition to confrontational protests and demonstrations, which many members of the public may not be able to join, participate, and engage in, proximate imageries and narratives of effective climate actions must be highlighted. These include involving in prefigurative energy solutions in communities, towns, and cities that demonstrate ethically preferable alternatives to the fossil fuel regime (Delina et al. 2014; North 2011; Bulkeley and Betsill 2003; cf. Fisher and McInerney 2012). SAE21 supports this approach, suggesting the "building [of] a better-integrated, more resilient local community that shares a strong set of environmental values and supports efforts to put them into action." With the proximity of counties, school boards, cities, and the like to the target audience of climate actions and their intimate involvement in everyday life, campaigners are in unparalleled positions in responding to people's needs and dreams, while, at the same time, advancing local climate actions.

---

[3]These vested interests comprise politicians, corporations, media, and other groups, who block climate legislations in Parliaments, lobby for continued fossil fuel extraction, undermine climate science, and spread disinformation in the media and public forums (Diesendorf 2009).

Since emissions from fossil-based technologies are one of the principal drivers of climate change, the climate action movement could also focus on promoting possibilities of energy transitions. Most of the technological options to transition to 80–100% renewable electricity systems with greater energy efficiency and public transport, for example, are already demonstrated and commercially available (see Delina 2016). These transitions can be accomplished by unleashing and celebrating the resources and skills of ordinary citizens in communities where low-carbon economies can be built from the bottom-up through many nimble community-owned strategies and approaches (e.g. Morris and Jungjohann 2016). Localised and tailored strategies that respond to the diversity of the many groups can motivate this type of prefigurative activism. At the same time, this kind of motivation becomes relevant to many since it can respond to what people personally care about.

Smaller alternatives provided by community-oriented climate actions have high promise of success (Manning and Reinecke 2016), although, of course, they also may fail. These small wins can be focused on smaller scale independent projects with attainable and measurable objectives, and can be participated by individuals. The example provided by decentralised community renewable energy production helps to enable communities to act as citizens, not just consumers. In these spaces of innovation, collective action is organised and articulated (Islar and Busch 2016), and could shape not only local but even national energy futures (Delina 2018; cf. Morris and Jungjohann 2016). To increase the level of success, these small-scale climate actions must multiply. As Morris and Jungjohann (2016) demonstrate in the case of the German *Energiewende*, supportive policy is necessary to shore them up. A networked approach, whereby small community energy systems are linked across a larger system via micro-grids, also appears to be essential. Linkages, however, are not only a province of connecting technology systems such as in microgrids, building stronger social networks across community groups also must be fostered and supported (see Chapter 5).

Divestments from fossil-based holdings, a type of shareholder action, are another example of winnable strategies with micro-level application that can be scaled in a networked approach. Initiated by students at a handful of US colleges in 2011, this strategy mimics the divestment campaign against the South African apartheid. Divestment builds a broad-based movement because everyone, everywhere has something to divest: from pension funds to personal investments. Ever since, many groups worldwide have embraced this strategy. In Australia, universities such as Australian National University and the University of Sydney are among those who divested their fossil fuel holdings, while researchers at the University of New South Wales, University of Queensland, and University of Melbourne have issued letters demanding their universities sell endowment investments in companies involved in fossil fuel extraction or processing (Hannam 2014). A number of religious groups have also divested from fossil fuel companies, such as the United Church of Christ in the USA (Raushenbush 2013), some branches of the Uniting Church of Australia (Australian Broadcasting Corporation Environment 2013), and the Church of Sweden (Bertini 2014). By 2014, more than 600 individuals, local and city governments, educational and healthcare institutions, and a number of non-government organisations have been divesting, collectively, more than US$50 billion of their assets (Arabella Advisors 2014). By 2017, 718 institutions and more than 58,000 individuals comprising US$5.45 trillion in investment dollars have divested from fossil fuels (Maxmin 2017).

Other small wins that can be achieved in localised solutions are by means of directing groups' purchasing power to move markets towards more climate-friendly products. A growing number of government departments, colleges and universities, NGOs, and companies have already established formal purchasing policies that encourage or require staff responsible for spending decisions to choose sustainable, greener options available. Industries could also embark on improving energy efficiency through cleaner production and reviewing their consumption profiles. Municipalities and cities also can influence climate actions through leading by example. They can adopt green public procurement policies and energy efficiency strategies in their buildings.

These diversified climate actions, however, are not limited towards technological change. Other equally important strategic areas must be embedded in these face-to-face heterogeneous climate actions. Some of these actions could include: weaning away from unrestricted consumption (Agyeman 2008); envisaging economic futures away from growth-led paradigms (Schneider et al. 2010), and questioning the 'privatising profit, socialising risk' maxim such as by campaigning towards energy market transformation (Pearse 2016) and reduced work hours (Schor 2010); and supporting alternative approaches for imagining what good life could mean such as the Buen Vivir where it is understood that the good of the community comes before that of the individual (Gudynas 2011).

In closing, relationships matter in converting talk into climate actions. The public comprehends campaigns based on the frames used, the narratives chosen, the tools employed, and the figures and voices that convey those messages. Physical presence remains important. Face-to-face meetings and gatherings present opportunity campaigners for climate actions must maximise. All these interventions require personal touches—and establishing honest and truthful relationships. Getting the right message across at the right time and at the right place is paramount in the strategies for mobilising and engaging people for climate actions. The action orientation of these conversations is vital but there is no guarantee that by focusing the message of mobilisation with action orientation alone would yield rapid and successful results. Such realisation is also important in a campaigner's portfolio. SAE33 agrees:

> I think that building social capital and a sense of belonging may work to slowly change people's beliefs/attitudes to climate change and other environmental issues...but it's a slow approach, and attitudes don't change overnight...(we) need to strengthen people's intrinsic values...unfortunately, though I haven't come across practical ways of achieving this, short of 'being the change you wish to see' and hoping that over time people will come on board and share your values/behaviours.

References

Agyeman, J. (2008). Toward a 'just' sustainability? *Continuum: Journal of Media & Cultural Studies, 22*, 751–756.

Arabella Advisors. (2014). *Measuring the global fossil fuel divestment movement.* http://bit.ly/1reDL6Q.

Assadourian, E. (2013). Building an enduring environmental movement. In E. Assadourian & T. Prugh (Project Directors), *State of the World 2013: Is Sustainability Still Possible?* (pp. 292–303). Washington, DC; Covelo, CA; and London, UK: Island Press.

Australian Broadcasting Corporation (ABC) Environment. (2013, 1 August). 'Unitarian church divests from fossil fuels.' *ABC News*, http://ab.co/1lIDjJZ.

Bertini, I. (2014, September 23). Church of Sweden completed fossil fuels divestment as movement doubles in one year. *Blue & Green Tomorrow.* http://bit.ly/1wLlKyM.

Bulkeley, H., & Betsill, M. (2003). *Cities and Climate Change: Urban Sustainability and Global Governance.* London, UK: Routledge.

Delina, L. (2016). *Strategies for Rapid Climate Mitigation: Wartime Mobilisation as Model for Action?* Oxon, UK: Routledge.

Delina, L. (2018). Can energy democracy thrive in a non-democracy? *Frontiers in Environmental Science, 6,* 5.

Delina, L., Diesendorf, M., & Merson, J. (2014). Strengthening the climate action movement strategies from histories. *Carbon Management, 5,* 397–409.

Diesendorf, M. (2009). *Climate Action: A Campaign Manual for Greenhouse Solutions.* Sydney, Australia: UNSW Press.

Fernandez, R. M., & MacAdam, D. (1988). Social networks and social movements: Multiorganizational fields and recruitment to Mississippi Freedom Summer. *Sociological Forum, 3,* 357–382.

Fisher, D. R., & McInerney, P.-B. (2012). The limits of networks in social movement retention: On canvassers and their careers. *Mobilization: An International Journal, 17,* 109–128.

Gudynas, E. (2011). Buen Vivir: Germinando alternativas al desarrollo. *América Latina en movimiento, 462,* 1–20.

Hannam, P. (2014, October 20). More Australian universities come under pressure to divest from fossil fuels. *The Sydney Morning Herald.* http://bit.ly/1qLBZI3.

Islar, M., & Busch, H. (2016). "We are not in this to save the polar bears!"— The link between community renewable energy development and ecological citizenship. *Innovation: The European Journal of Social Science Research, 29,* 303–319.

Jasper, J. M., & Poulsen, J. D. (1995). Recruiting strangers and friends: Moral shocks and social networks in animal rights and antinuclear protests. *Social Problems, 42,* 493–512.

Kirby, A. (2008). *A UN Guide to Climate Neutrality.* Malta: United Nations Environment Programme and GRID-Arendal.

Klandermans, B., & Oegema, D. (1987). Potentials, networks, motivations, and barriers: Steps towards participation in social movements. *American Sociological Review, 52,* 519–531.

Læssoe, J. (2007). Participation and sustainable development: The post-ecologist transformation of citizen involvement in Denmark. *Environmental Politics, 16,* 231–250.

Manning, S., & Reinecke, J. (2016). A modular governance architecture in-the-making: How transnational standard-setters govern sustainability transitions. *Research Policy, 45,* 618–633.

Maxmin, C. (2017, June 5). How Harvard divestment was won. *The Nation.*

Morris, C., & Jungjohann, A. (2016). *Energy Democracy: Germany's Energiewende to Renewables.* London, UK: Palgrave Macmillan.

Moser, SC., & Dilling, L. (Eds.). (2007). *Creating a Climate for Change: Communicating Climate Change and Facilitating Social Change*. Cambridge, UK: Cambridge University Press.

North, P. (2011). The politics of climate activism in the UK: A social movement analysis. *Environment and Planning A, 43*, 1581–1598.

Opp, K.-D., & Gern, C. (1993). Dissident groups, personal networks, and spontaneous cooperation: The East-German Revolution of 1989. *American Sociological Review, 58*, 659–680.

Pearse, R. (2016). Moving targets: Carbon pricing, energy markets, and social movements in Australia. *Environmental Politics, 25*, 1079–1101.

Raushenbush, P. B. (2013, July 2). Fossil fuel divestment strategy passes at United Church of Christ Convention (UCC). *The Huffington Post*. http://huff.to/1xVzYfC.

Saunders, C. (2008). Double-edged swords? Collective identity and solidarity in the environmental movement. *The British Journal of Sociology, 59*, 227–253.

Schneider, F., Kallis, G., & Martinez-Alier, J. (2010). Crisis or opportunity? Economic degrowth for social equity and ecological sustainability. *Journal of Cleaner Production, 18*, 511–518.

Schor, J. B. (2010). *Plenitude: The New Economics of True Wealth*. New York, USA: The Penguin Press.

Snow, D. A. (2013). Framing and social movements. In D. A. Snow, D. della Porta, B. Klandermans, & D. McAdam (Eds.), *The Wiley-Blackwell Encyclopedia of Social and Political Movements*. Malden, MA: Blackwell.

Snow, D. A., Rochford, B., Worden, S. K., & Benford, R. D. (1986). Frame alignment processes, micromobilization, and movement participation. *American Sociological Review, 51*, 464–481.

Vala, C. T., & O'Brien, K. J. (2007). Attraction without networks: Recruiting strangers to unregistered protestantism in China. *Mobilization, 12*, 79–94.

CHAPTER 3

Messaging

Abstract Psychological stimuli can change people's attitudes towards social change; yet their use can also open up tensions that can easily backfire especially when not well managed. The strategic use of these stimuli in social movements often depends upon how skillfully campaigners could orient their messages vis-à-vis the strongly held values and moral concerns of their respective audiences. While evoking moral concerns in campaigns can be effective in mobilisation, it, too, can easily backfire. This chapter asks how can future messages be adapted to the moral compasses of particular audiences and with dexterity, what are the pitfalls of morally charged messages for climate actions, and how could activists—using examples from other mobilisations—address these through value-based messaging.

Keywords Climate psychology · Climate stimuli · Values · Morality

I do this by 'pointing out how an issue affects each person – their health, their wallet, their children, the things they value and enjoy. Everyone is an environmentalist when the issue affects them.'
—Respondent from a U.S.-based environmental group formed in 2003 helping individuals with environmental problems in their respective neighbourhoods. The group has also taken part in state-wide environmental campaigns against fracking.

© The Author(s) 2019   29
L. L. Delina, *Climate Actions*,
https://doi.org/10.1007/978-3-319-91884-6_3

How we act toward the various objects of orientation that populate our daily lives depends on how they are framed (Snow 2013). Meanings do not automatically or naturally attach themselves to the objects, events, or experiences we encounter; rather, they arise through interpretive processes mediated by culture. Clearly, culture has effects on how we view ourselves and how others perceived us. Frames contribute to this interpretive work by focusing attention (highlighting what is relevant and irrelevant), articulating (tying elements so that one story vs. another is told), and transforming (mobilising grievances) (Snow 2013).

A branch of the psychology literature suggests people tend to react with passion and determination following psychological tipping points such as their keen realisation that the values they strongly hold are being violated (Markowitz and Shariff 2012; Swim et al. 2011). These stimuli appear to have caused large-scale activism to ignite and were strategically used by social movements as symbols for strengthening their respective struggles, as well as mechanisms for mobilising support (Nepstad 2011; Delina et al. 2014). Most survey respondents agree that shocking events could produce 'effective' campaigns ($N = 38$, median $= 4$, Interquartile Range (IQR) $= 0$[1]), with nine or 24% even saying that invoking such an incident or event could be 'very effective' in mobilising people. Without a shocking event to highlight campaigns, ten respondents or 21% state that public attention would be hard to achieve.

Many in the climate action movement have already (and continue to) hinged their strategies using vocabularies related to shocking events— including even those yet projected to occur. They have been employing images directly or indirectly related to past, present, and future events to stimulate attention and engagement. Others even bank on exogenous shocking events in the form of extreme weather events to spur people onto action. SAE35[2] contends:

> At this point it seems that highlighting recent unpredictable weather and extreme weather events in the context of climate change has gotten people's attention, and explaining the scientific consensus that irreversible and catastrophic global warming is likely within this century if there is no coordinated global effort to slash greenhouse gas emissions that begins in

[1] See Appendix B for notes on statistical treatment and reporting.
[2] See Table C.1 for the description of the social action groups.

the next decade is enough to make those we've spoken to support our call for a national climate mobilization.

Campaigners are also using many symbols in climate action mobilisation: images of a lone, emasculated polar bear stranded on a breaking ice; a submerged coastal city; a graph of fast-rising emissions; etc. They hope that with these images they can attract people's attention and appeal to the moral positions to protect endangered species and future generations of humans and non-humans. However, people's perceptions of these types of climate symbolism and vocabularies vary.

While burden framing can grab some attention and may mobilise some to adopt climate actions, the focus on spatially and temporally distant events, technical language, and fear appeals can easily backfire and even disengage some audiences (Weber 2010; Moser 2007a, c). In some instances, this approach is even perceived as a form of manipulation (Lorenzoni et al. 2007; O'Neill and Nicholson-Cole 2009). The reasons for this are manifold but the behavioural and brain sciences explain that the human moral judgment system is poorly equipped to identify future, large-scale, long-term hazards such as climate impacts (Swim et al. 2011).

The evolutionary history of our species simply impedes our capacity to react today on the future ramifications of our historical and present actions (Gifford 2011). Psychological repercussions, such as denial and apathy, also render burden framing of limited value in climate action mobilisation (American Psychological Association 2009; Center for Research on Environmental Decisions 2009). The problem with too much reliance on burden frames to catapult climate actions is that 'the [climate action] movement may not be granted similar pivot events that could highlight the moral basis for effective climate action' (Delina et al. 2014: 402). Such extreme events could also occur too late or develop too gradually to be effectively used as narratives in driving effective climate actions. Even if an extreme event occurs, there is no guarantee it could also gain salience. SAJ01 notes: "While a critical incident tends to generate interest in existing campaigns, they are not particularly helpful for new ones—the interest simply burns out as soon as the issue had disappeared from the news cycle." Some climate actions, such as a sustainable energy transitions and transformations in the agriculture and forestry industries, would also take at least several decades to accomplish. One of the biggest challenges faced by climate action campaigners, thus, is creating a strong sense of the need for "doing what is right" urgently across all sectors of society, given the spatial and temporal distance of the large-scale effects of climate change and the benefits of climate actions now.

The gap between climate actions and their truly measurable effects, thus, betrays campaigners with an argument for the 'now.' Since climate actions today will largely be for the benefit of future generations, the climate action movement must develop strategies that focus other forms of benefits of mitigation for the present. There is no easy solution to that end, but motivating people can be best revolved around people's personal, lived experiences (Fisher and McInerney 2012). In gist, climate actions must be about translating 'the mundane politics of everyday life, into a directly embodied political process of movement mobilisation for a genuine strategy for transformation' (Rosewarne et al. 2014: 17).

Invoking proximate rationales must be broadly consistent with the audience's personal aspirations, desired social identity, and cultural biases (Fisher and McInerney 2012; Weber 2010; Segnit and Ereaut 2007; Markowitz and Shariff 2012; Swim et al. 2011; Haidt 2007). Orienting and connecting campaigns on values the audience holds dear is key (American Psychological Association (APA) 2009; Lorenzoni et al. 2007). Indeed, it seems paradoxical that climate action messages must be conveyed to some audiences not with scientific facts to convince them but with the values they hold dear to give the facts a fighting chance (Mooney 2011; cf. Kahan et al. 2011).

Most survey respondents agree that values-based campaigns have been effective in gathering support ($N=45$, median$=4$, IQR$=1$). SAE24 notes: "Reframing questions to take into account 'green fatigue' such as do you support renewable energy instead of do you believe in climate change [can be effective]." SAE04 does this by "pointing out how an issue affects each person – their health, their wallet, their children, the things they value and enjoy." This approach also rings true to audiences in many fast urbanising urban areas in developing countries, where climate change may not be the most effective frame but pollution from coal and its effects on their health really resonate.

In developing these proximate frames, campaigners must clearly identify personal values that people hold dearly. Since this exercise requires a thorough reading of what the audience values, it entails the investment of time and effort for the campaigners. Aligning climate actions with representations brought about by proximate, local icons that this audience can easily relate to can assist in frame development. This way, the audience can really care about climate actions and can empathise with the campaigners. These tools should be preferred over distant and complex imageries and symbols (O'Neill and Hulme 2009; Weber 2006). Among the many examples of these framing tools and symbols are local sustainable solutions, such as community energy, local public transport, building efficiency, and sustainable consumption.

Value-based mobilisations, however, need to extend beyond what can be directly developed out of people's needs, aspirations, priorities, dreams, and hopes. When triggering outward-oriented climate actions, campaigners may found the importance of moral benchmarking, particularly by stressing how current regimes violate deeply held moral values. Psychological studies have already established that people realise the inevitability of change following their experience of value violations firsthand (Center for Research on Environmental Decisions 2009; Weber 2010; American Psychological Association 2009). When people see their strong sense of right and wrong, and the values they hold precious, are transgressed, large-scale activism tends to be ignited (Moyer et al. 2001). In many past social movements, campaigners strategically used these violations to mobilise public support (Delina et al. 2014). Some historical examples include the arrest of Rosa Parks (which ignited the modern civil rights movement); the assassination of Ninoy Aquino (which spurred the Philippine People Power Revolution); and the violent dispersal of peaceful demonstrators in Dharasana (which galvanised international recognition of India's struggle for independence) (cf. Nepstad 2011).

Within the climate action movement, the moral basis for strong climate actions exists. Climate change has already been extensively framed as a matter of right and wrong; and, for some time already, such framing has been a part of the recipe of many climate actions. Bill McKibben, one of the most prominent and influential voices in the movement, writes: "[The] more carefully you do the math, the more thoroughly you realise that [climate change], at bottom, [is] a moral issue" (McKibben 2012: 5). In the field, SAE04, describes how their group connects morality with their campaign against natural gas.

[Using] Strong science showing right from wrong [is vital]. [For example, the] IPCC says methane is 86 times more potent than CO_2, this pipeline would more than double Massachusetts CO_2 equivalent emissions at a time when we are the leading state in reducing those emissions, at the same time our economy has been growing, especially in the green jobs sector. Adding a new fracked [sic] gas pipeline is [therefore] a moral issue.

SAO07 also notes:

> You need to frame an issue to make it emotive, memorable, with a clear wrong, and a convincing solution that requires the individual to take action. It needs to be simple enough to be easy to communicate quickly to busy people with no prior interest, but the quick and simple version needs to avoid distortions.

Calls to avoid injustice now and beyond, and to avoid unacceptable impacts on people today and the future generations are two moral values, according to survey respondents, that can be simultaneously invoked to bring the focus on climate solutions today. This moral imperative is key in setting the normative definition of climate actions—the way the world *ought* to be (Dean Moore and Nelson 2013)—and in understanding its temporal terms, both intergenerational and immediate (Rosewarne et al. 2014: 91). Social justice is a common thread among contemporary values that respondents indicate as 'important' in mobilisation ($N=46$, median$=5$, IQR$=1$).

Justice can be conceptualised in myriad ways, but a common understanding of the basic needs, rights, and political processes to be engaged and protected to as we do climate actions is central to this argument. This approach incorporates justice with fair distributions (what the concerns are), political and social recognition (who it affects), and procedural inclusion (how it could be mediated or remediated) (Sovacool and Dworkin 2014; Jenkins et al. 2016). Survey respondents hold that basic human rights with special protection for rights of children and the minorities in society as an important value ($N=43$, median$=4$, IQR$=2$). Justice also pertains beyond the present generation of human beings to include the unborn and the future generation. Twenty-five respondents or 54% state that invoking intergenerational equity in campaigns is 'very important,' while 11 or 24% declare it 'important.' For most respondents, justice even extends to non-humans, where most respondents ($N=46$, median$=5$, IQR$=1$), for example, state the importance of conserving biodiversity along with ecological integrity in campaigning. Twenty-eight respondents or 61% declare that anchoring their campaigns to this value is 'very important,' while 13 or 28% claim it as 'important.'

Within the concept of justice also lies procedural inclusion, meaning that climate actions must be clear and explicit and done with active public engagement on the different values, discourses, and potential courses of action. Distributional justice responds to this. What are the distributional burdens of climate actions, and how do we address them are but two of the many questions related to the distributional conception of justice. Democratically organised and mobilised climate actions, thus, are key in embedding the voices of the traditionally ignored and/or misrepresented sections of society. This includes understanding and evaluating who the ignored actors are and how should they be recognised. Recognition-based justice, however, must move towards ensuring procedural justice: Are the processes fair and what new exercises or processes can be developed to ensure fair processes. Democratising climate actions mean the creation of new (or strengthening existing) systems, processes, and arrangements, including participatory decision-making that consequentially led to stronger, inclusive public engagement in climate actions. Climate actions of this kind demand policy processes conducted in ways that allow for the recognition and representation of interests, values, and reflections of community members in a respectful space. Protecting and extending democracy, therefore, is an ideal worthy of being prized, and, therefore, fought for by the climate action movement.

One important way to ensure democracy within climate actions is to protect broad democratic ideals—including the right to dissent, which, even in established democracies, is under constant threat. For example, in the supposedly democratic societies of Australia and the USA, democracy has been relentlessly repressed in formal institutions. The ways by which power-holders continue to systematically embed structural impediments in institutional arrangements and governmental policies aimed at curtailing dissent are key examples. For instance, the rights of citizens to peacefully and nonviolently assemble and demonstrate have been threatened in the state of Victoria in Australia. In 2014, the Parliament of Victoria (2014) passed a legislation amending the *1966 Summary Offences Act*. With this law, Victorians protesting about social or environmental issues are now subject to police powers that could order them to further 'move-on' from a public place and/or face arrest, and in some circumstances to detain them in custody. The law could effectively deny individual and group's right to assemble, picket, and conduct political rallies in the state. The day before the third and final reading of the bill in Parliament, the Victorian Greens issued this statement:

> The bill is an absolute assault on the democratic right of Victorians to protest...about issues of concern to them...Police already have sufficient powers currently under the [1966] Act...the main changes to the Act... are completely unnecessary and go to the democratic right of Victorians to protest... In protesting...they may in fact be inconveniencing other members of the public or business or other organisations, and that is the point of the protest...This legislation has popped out of nowhere and is designed to shut down community protest and activism.... (Pennicuik 2014)

This legislation in Victoria is only one among the many attempts to criminalise protests in Australia and internationally. Other examples have been cropping up in other Australian states. The Hodgman Liberal opposition in the state of Tasmania, for example, had proposed AU$10,000 fines for environmental protesters who enter or impede access to workplaces, with mandatory jail sentences for second offences and property damage (Billings 2013). The Newman Liberal National government in the state of Queensland proposed a legislation providing police with the power to arrest and imprison any three or more members of 'unlawful association' who are assembled in public (McNamara 2014). In Canada, the country's domestic spy agency increased its surveillance on activists opposed to the Northern Gateway pipeline project on 'national security' grounds (Millar 2013). This 8 billion Canadian dollar project plans to transport oil from Alberta tar sands to the British Columbia coast where it can then be shipped to global markets. In the USA, extensive evidence shows how some of the world's largest corporations collaborated with the national security apparatus of the federal government to spy on activists, whistleblowers, and non-profit groups under the rubric of 'anti-terrorism' (Ruskin 2013). These repressive moves from governments and some corporations subvert democracy. Reining in the forces of vested interests, which in many ways are also threatening inherent human values including democracy, therefore, is one of the many moral battles that the climate action movement must continue to wage.

Outward-oriented activists are often reminded of the value of nonviolence in its campaigns at all times. Nonviolence has been proven to be an important characteristic of the transformative character of past social action movements (Delina et al. 2014) and many respondents also agree that nonviolence remains an important parcel of social actions ($N=42$, median $=5$, IQR $=1$). Evoking images of nonviolent social actions alone, historical mobilisations suggest, could help in making some disaffected elites to quickly switch sides and many ordinary citizens to participate (Nepstad 2011). Twenty-six respondents or 62% declare that anchoring campaigns to nonviolence was 'very important,' while 10 or 24% state that this has been 'important.' But nonviolence is only one of the many virtues that campaigners for climate actions must exemplify. The ways they are framed as messengers of climate actions also matter.

The consistency between claims and actions and the credibility of climate action articulators indeed affect the credibility of the climate action message (cf. Snow 2013). Living by example appears to be vital for effective campaigns. Charles Tilly (1995, 2002, 2008) associates this part of social action with 'worthiness.' Most respondents agree that their campaigns become more 'effective' if people see them as somebody who exhibits 'good' values and moral virtues ($N=46$, median $=5$, IQR $=1$). Twenty-seven respondents or 59% claim that 'campaigning by example' is 'very effective,' while 18 or 39% consider it as an 'effective' approach. Some respondents mention integrity (SAE03, SAE37), honesty (SAE37), and humility (SAE03) as some of these essential values.

Of course the quality of the message, not just the messenger's, is important. Some climate action organisations have already recognised this as an essential and have set training platforms such as Climate Reality Project's Climate Leadership Corps. In these platforms, campaigners can be skilled in climate action messaging including storytelling, public speaking, social media networking and media engagement. With information and communication technologies, 'skilling' activities can now be done 24/7 without needing physical instructors and classrooms. Massive open online courses, instructional videos, and training booklets can, indeed, now be produced and widely distributed, thus minimising the costs of face-to-face instructions. (Although of course the latter is preferable given its many advantages beyond education, including networking opportunities among like-minded and goal-oriented individuals (Yagatich et al. 2018; also see Chapter 1)). This mode of skilling also allows easy campaign replication. The idea is to design training programmes that lead to a snowballing effect where many other climate action campaigners can replicate similar trainings, workshops, and meetings in other communities, households, schools, workplaces, clubs, and associations. Campaigners, however, must be reminded to always locate the issues they discuss within appropriate and, as much as possible, proximate frames (see Chapter 1); otherwise, the importance and credibility of the message can be weakened or even jeopardised (FrameWorks

Institute 2002). Building a legion of campaigners that can relate to several possible framings—and to scale these exercises—is a necessary campaign strategy for the climate action movement.

Ideally, campaigners should be able to choose frames that resonate best with the intended audience. Our survey indicates that having campaigners who resonate with particular audiences is important. The logic suggested here is that if the target audience is a crowd of teenagers, for example, popular celebrities may be able to generate strong engagement (Boykoff and Goodman 2009). To most respondents, having prominent people becoming involved and engaged in their campaigns is indeed a condition for 'effective' campaigning ($N=41$, median$=4$, IQR$=1$). The absence of prominent figures, according to twenty-four respondents or 51%, poses a barrier. Twenty-seven respondents or 66% declare that the presence of prominent people in their activities resulted to more 'effective' campaigns, while 11 or 27% state having these people around could be a 'very effective' strategy. SAE33 describes this approach:

> Engaging with people in positions of influence [is one of our group's strategies]. Community leaders (i.e. well-liked, well-respected members of the community) can influence the beliefs/attitudes of others. Hence, we try to achieve social change by encouraging business leaders, political leaders, church leaders, emergency personnel and the like to publicly state their concerns about climate change and the need for stronger action. We believe that media stories featuring these people are more likely to accomplish social change than repeated stories featuring local climate activists. In short, we believe it's not just the climate message, but [also] the climate messenger that's important in achieving social change.

An example can be plucked from the world of sports. For many people, footballers are their influencers. Thus, when the British club Ipswich Town chose to participate in climate actions, a lot of people paid attention. The club determined that they produce 3200 tonnes of carbon dioxide per season and decided to offset this by asking their supporters to make specific pledges to save energy. The fans were encouraged to reach the target by committing themselves to take simple steps like using public transport and high-efficiency light bulbs, while some of the players turned to car-pooling. Along the same vein, religious people may be the best ones to communicate climate actions especially when they are framed as a moral issue (Wardekker et al. 2009). The civil rights movement, where many of the voices came from pastors of religious denominations, illustrates this approach (Houck and Dixon 2006; Hall 2005; Marx 1967). In communicating climate actions, conservatives are more likely to embrace climate actions if the message comes to them through their religious leaders, who can set the issue in the context of different values than from which environmentalists or scientists often argue (cf. Kahan et al. 2011).

Although preferable, having volunteer campaigners from all possible frames and locations may be anything but impossible. In the absence of celebrities, campaigners have to effectively and efficiently use whatever resources they have. Preparing campaigners to become effective communicators and facilitators regardless of audience and location is, thus, essential. For instance, young adults who are active in the climate action movement can become effective campaign voices in youth groups (Isham and Waage 2007). The Australian Youth Climate Coalition (AYCC), founded in 2007, provides an example. AYCC trains young campaigners in helping young people take action in their own communities, schools, and universities, and mobilise them for national climate actions (AYCC 2013). What is key to extend in this approach is to have campaigners, who although are from different frames, can still be listened to as long as they are trustworthy and credible (cf. Fessenden-Raden et al. 1987). SAE30 suggests this level of engagement is attained when campaigners "always speak the truth and have courage of conviction." To that end, trust is precious that climate action campaigners must cultivate this with their audiences (Buys et al. 2014; Marx et al. 2007). Building trust, however, takes time, but once established, studies suggest people perceive campaigners as the right people conveying the right message and, therefore, are welcomed (Buys et al. 2014; Agyeman et al. 2007; Cialdini 1993).

In closing, climate action messages must be framed based on the value characteristics of the intended audience. Optimistic messages are key, but without sacrificing fact-based science. A key caveat: dreaming is not doing. An optimistic goal may be crippling when it comes to working towards it. Rosy messages can zap the motivation to mobilise thus leaving us with lofty ideas that may never reach fruition (Oettingen 2015). The emphasis on moral messages and values-based framing, thus, remains an important strategy. Vocabularies and symbols are necessary to prod people into climate actions. These vocabularies, however, must be spatially and temporally proximate to the recipients of the message. They have to ring well with the audience's priorities, needs, dreams, and hopes. If done correctly, these symbols and vocabularies of climate actions can help overcome the challenges and risks associated with collective climate actions. While messengers must be trustworthy when framing these messages, it remains essential that they also balance the message pertaining to the severity of climate impacts by accenting the urgency of taking action now, the efficacy of collective climate actions, and the moral priority of doing so.

REFERENCES

Agyeman, J., Doppelt, B., Lynn, K., & Hatic, H. (2007). The climate-justice link: Communicating risk with low-income and minority audiences. In S. C. Moser & L. Dilling (Eds.), *Creating a Climate for Change: Communicating Climate Change and Facilitating Social Change* (pp. 119–138). Cambridge, UK: Cambridge University Press.

American Psychological Association (APA). (2009). *Psychology and Global Climate Change: Addressing a Multi-faceted Phenomenon and Set of Challenges: A Report by the Task Force on the Interface Between Psychology and Global Climate Change.* Washington, DC: APA.

Australian Youth Climate Coalition (AYCC). (2013). *About AYCC.* http://bit. ly/1pm8daW.

Billings, P. (2013, December 8). Protesters in Hodgman's sights. *The Examiner.* http://bit.ly/1d5a931.

Boykoff, M. T., & Goodman, M. K. (2009). Conspicuous redemption? Reflections on the promises and perils of the 'celebritization' of climate Change. *Geoforum, 40,* 395–406.

Buys, L., Aird, R., Van Megen, K., Miller, E., & Sommerfeld, J. (2014). Perceptions of climate change and trust in information providers in rural Australia. *Public Understanding of Science, 23,* 170–188.

Center for Research on Environmental Decisions (CRED). (2009). *The Psychology of Climate Change Communication: A Guide for Scientists, Journalists, Educators, Political Aides, and the Interested Public.* New York, NY: CRED, Columbia University.

Cialdini, R. B. (1993). *Influence: The Psychology of Persuasion* (2nd Rev. ed.). New York: Quill-William Morrow.

Dean Moore, K., & Nelson, M. P. (2013). Moving toward a global moral consensus on environmental action. In *State of the World 2013: Is Sustainability Still Possible?* Washington, DC; Covelo; and London, UK: Island Press.

Delina, L., Diesendorf, M., & Merson, J. (2014). Strengthening the climate action movement strategies from histories. *Carbon Management, 5,* 397–409.

Fessenden-Raden, J., Fitchen, J. M., & Heath, J. S. (1987). Providing risk information in communities: Factors influencing what is heard and accepted. *Science, Technology, and Human Values, 12,* 94–101.

Fisher, D. R., & McInerney, P.-B. (2012). The limits of networks in social movement retention: On canvassers and their careers. *Mobilization: An International Journal, 17,* 109–128.

FrameWorks Institute. (2002). *Framing Public Issue.* Washington, DC: FrameWorks Institute.

Gifford, R. (2011). The dragons of inaction: Psychological barriers that limit climate change mitigation and adaptation. *American Psychology, 66,* 290–302.

Haidt, J. (2007). The new synthesis on moral psychology. *Science, 316,* 998–1002.

Hall, J. D. (2005). The long civil rights movement and the political uses of the past. *Journal of American History, 91,* 1233–1263.

Houck, D. W., & Dixon, D. E. (Eds.). (2006). *Rhetoric, Religion and the Civil Rights Movement, 1954–1965.* Waco, TX: Baylor University Press.

Isham, J., & Waage, S. (Eds.). (2007). *Ignition: What You Can Do to Fight Global Warming and Spark a Movement.* Washington, DC: Island Press.

Jenkins, K., McCauley, D., Heffron, R., Stephan, H., & Rehner, R. (2016). Energy justice: A conceptual review. *Energy Research & Social Science, 11,* 174–182.

Kahan, D. M., Jenkins-Smith, H., & Braman, D. (2011). Cultural cognition of scientific consensus. *Journal of Risk Research, 14,* 147–174.

Lorenzoni, I., Nicholson-Cole, S., & Whitmarsh, L. (2007). Barriers perceived to engaging with climate change among the UK public and their policy implications. *Global Environmental Change, 17,* 445–459.

Markowitz, E. M., & Shariff, A. F. (2012). Climate change and moral judgment. *Nature Climate Change, 2,* 243–247.

Marx, G. T. (1967). Opiate or inspiration of civil rights militancy among Negroes? *American Sociological Review, 32,* 64–72.

Marx, S. M., Weber, E. U., Orlove, B. S., Leiserowitz, A., Krantz, D. H., Roncoli, C., et al. (2007). Communication and mental processes: Experiential and analytic processing of uncertain climate information. *Global Environmental Change, 17,* 47–48.

McKibben, B. (2012, July 19). Global warming's terrifying new math. *Rolling Stone.* http://rol.st/LuRoru.

McNamara, M. (2014, January 29). Criminalising activism a growing trend. *Echo Net Daily.* http://bit.ly/100GuYE.

Millar, M. (2013, November 19). Harper government's extensive spying on anti-oil sands group revealed in FOIs. *Vancouver Observer.* http://Bit. Ly/1ddnsqs.

Mooney, C. (2011, May/June). The science of why we don't believe science. *Mother Jones.* http://bit.ly/2Ez9egt.

Moser, S. C. (2007a). Communication strategies to mobilize the climate movement. In J. Isham & S. Waage (Eds.), *Ignition: What You Can Do to Fight Global Warming and Spark a Movement* (pp. 73–95). Washington, DC: Island Press.

Moser, S. C. (2007c). More bad news: The risk of neglecting emotional responses to climate change information. In S. C. Moser & L. Dilling (Eds.), *Creating a Climate for Change: Communicating Climate Change and Facilitating Social Change* (pp. 64–80). Cambridge, UK: Cambridge University Press.

Moyer, B., McAllister, J., Finley, M. L., & Soifer, S. (2001). *Doing Democracy: The MAP Model for Organizing Social Movements.* Gabriola Island, BC: New Society Publishers.

Nepstad, S. E. (2011). *Nonviolent Revolutions: Civil Resistance in the Late 20th Century.* New York: Oxford University Press.

O'Neill, S. J., & Hulme, M. (2009). An iconic approach for representing climate change. *Global Environmental Change, 19,* 402–410.

O'Neill, S. J., & Nicholson-Cole, S. (2009). "Fear won't do it": Promoting positive engagement with climate change through visual and iconic representations. *Science Communication, 30,* 355–379.

Oettingen, G. (2015). *Rethinking Positive Thinking: Inside the New Science of Motivation.* New York: Penguin.

Parliament of Victoria. (2014). *Summary Offences and Sentencing Amendment Act 2014.* Victoria, Australia. http://bit.ly/1rkMbvW.

Pennicuik, S. (2014, March 11). *On the Summary Offences and Sentencing Amendment Bill 2013.* Website of the Victorian Greens. http://bit.ly/1mA33x6.

Rosewarne, S., Goodman, J., & Pearse, R. (2014). *Climate Action Upsurge: The Ethnography of Climate Movement Politics.* Abingdon, Oxfordshire, UK, and New York: Routledge.

Ruskin, G. (2013). *Spooky Business: Corporate Espionage Against Nonprofit Organizations.* http://bit.ly/1DGhUeQ.

Segnit, N., & Ereaut, G. (2007). *Warm Words II: How the Climate Story Is Evolving and the Lessons We Can Learn for Encouraging Public Action.* London, UK: Institute for Public Policy Research.

Snow, D. A. (2013). Framing and social movements. In D. A. Snow, D. della Porta, B. Klandermans, & D. McAdam (Eds.), *The Wiley-Blackwell Encyclopedia of Social and Political Movements.* Malden, MA: Blackwell.

Sovacool, B. K., & Dworkin, M. H. (2014). *Global Energy Justice: Principles Problems and Practices.* Cambridge, UK: Cambridge University Press.

Swim, J. K., Stern, P. C., Doherty, T. J., Clayton, S., Reser, J. P., Weber, E. U., et al. (2011). Psychology's contributions to understanding and addressing global climate change. *American Psychologist, 66,* 241–250.

Tilly, C. (1995). *Popular Contention in Great Britain, 1754–1834.* Cambridge, MA: Harvard University Press.

Tilly, C. (2002). *Stories, Identities and Political Change.* New York: Rowman & Littlefield.

Tilly, C. (2008). *Contentious Performances.* Cambridge, UK: Cambridge University Press.

Wardekker, J. A., Petersena, A. C., & Sluijs, J. P. (2009). Ethics and public perception of climate change: Exploring the Christian voices in the US public debate. *Global Environmental Change, 19,* 512–521.

Weber, E. U. (2006). Experience-based and description-based perceptions of long-term risk: Why global warming does not scare us (yet). *Climatic Change, 77,* 103–120.

Weber, E. U. (2010). What shapes perceptions of climate change? *WIREs Climate Change, 1,* 332–342.

Yagatich, W., Galli Robertson, A. M., & Fisher, D. R. (2018). How local environmental stewardship diversifies democracy. *Local Environment, 23,* 431–447.

CHAPTER 4

Visioning

Abstract In the continuum that describes social movements, the most important phase is when a large majority of people starts to actively respond to the social issue by engaging in political actions and in changing their personal behaviours and mindsets. People could be provided with this new sense of collective identity and ownership when campaigners offer them a clear and unified regime alternative. This chapter asks how can climate action campaigners construct a new vision for a stable climate era, what are the complexities involved in these processes, what examples are blossoming elsewhere, how do they materialised, and how could these be used as visioning materials for climate actions, and, lastly, what are the weaknesses of this strategy as experienced in other social action campaigns.

Keywords Behavioural change · Vision · Community energy

> You need to frame an issue to make it emotive, memorable, with a clear wrong, and a convincing solution that requires the individual to take action. It needs to be simple enough to be easy to communicate quickly to busy people with no prior interest, but the quick and simple version needs to avoid distortions.
> —Respondent from a UK, faith-based, international development group formed in 1968 working on long-term development and disaster relief.

© The Author(s) 2019
L. L. Delina, *Climate Actions,*
https://doi.org/10.1007/978-3-319-91884-6_4

For a social cause to grow beyond local groups and communities, it must be able to propagate itself. Growing a social cause requires the transformation of citizens from mere participants and followers to active members of 'a collectively organised, self-directing, and highly engaged social change group' (Delina et al. 2014: 400). People must transform their prior understandings for the world to be seen differently from before. This process involves building an identity following an 'interactive and shared definition produced by several individuals (or groups at a more complex level)' (Melucchi 1989: 34). But it is key to note that identity-building is 'supposedly a *process* under continuous reflexive revision' (Saunders 2008: 230).

In the continuum describing social movements, identity-building is the most important stage (Diani 1992). When protesters were given a new sense of self-identity, historical mobilisations reveal, campaigns became stronger movements. People who started acting because they had taken ownership of the larger movement fuel these mobilisations. This new identity brings feelings and new experiences of solidarity, belongingness, harmony, and unanimity. Hence, individuals, though not personally acquainted but with which one nonetheless can share similar aspirations, can be mobilised together (Della Porta and Diani 2006; Melucci 1996; Diani 1995). During this stage, people actively respond to the social issue by involving themselves not only in political actions but also in changing their behaviour and mindset (Lorenzoni et al. 2007; Moser 2009). Survey respondents support this effort, underlining that when groups collaborate towards goals and clear demands, campaigns become more 'effective' ($N = 46$, median $= 4$, Interquartile range (IQR) $= 1^1$). Once strategies around identity-building are designed, people appear to become self-directing, and the chances of having successful campaigns are increased. Since this phase entails inculcating a new identity, it is also the most challenging.

One mechanism to build identity is to provide a clear regime alternative. Clarity is key when delivering a strong public performance (Tilly 1995, 2002, 2008) and highlighting a clear alternative to usher in a new identity has also been evident in past social mobilisations (Delina, Diesendorf and Merson 2014). SAE33 highlights that the "failure to develop a clear narrative on climate change that will engage and motivate

---

[1] See Appendix B for notes on statistical treatment and reporting.

people" leads to 'ineffective' campaigns. Indeed, greater engagement was secured following people's realisation that the movement could provide them an alternative to the status quo, a sense of community for its participants, and an example of virtues they want to own for themselves (Saunders 2008). Like master algorithms, clear visions of what the future ought to look like colours and constrains subsequent heterogeneous actions—the derivatives of a visioning exercise (Snow 2013). This kind of clarity of vision, however, is best applied to the group level rather than the larger movement as a whole (Saunders 2008).

The heterogeneity of group approaches, technologies, tools, and tactics for climate actions is a strong facet of the climate action movement; yet this plurality must be based on a positive vision (Schock 2005), or at least 'bridged' or linked despite the structural disconnections among individuals and groups (Snow et al. 1986; also see Chapter 5). While campaigners advance multiple futures, a vision, which 'may be broad but many groups can agree with, at least in principle' (Delina et al. 2014), is imperative to glue individual and group actions together.

At present, many climate action campaigns are planned, centred, and accomplished around what Peter North (2011) terms as 'outward-focused activism' through confrontational protests and demonstrations. While the movement must continue to strengthen campaigns of this type to target the worst obstructionist actors for effective climate actions, a focus on confrontational dissent alone remains a one-sided response. Equally important is for the movement to underline what North (2011) calls 'prefigurative activism' where as many people as possible are engaged in climate actions and solutions that focus on building regime alternatives. SAE35[2] shares: "[The] lack of concrete set of demands and actions to address grievances [is one of the barriers to getting greater public engagement]."

Realistic, in addition to clear, objectives are also essential in forging new identities. SAE36 observes: "Ensuring that our values, objectives, and messaging are realistic and are about achieving inclusivity with regards to membership, political views, etc. [has led to 'effective' campaigns]." This necessitates "providing people with concrete ways to take action" (SAE17) and "encouraging them to take practical steps" (SAE03) on their own (SAE13) and for their own development

---

[2] See Table C.1 for the description of the social action groups.

(SAE15). Some specific strategies to ensure this occurs include: "encouraging local communities to undertake environmentally aware projects such as local clean energy and energy efficiency" (SAE11); and "undertaking projects that capture community interest [such as] getting commitment to community targets, solar bulk buy schemes, and community owned solar farm" (SAE08). In constructing feelings of shared identities (Della Porta 2005) concerning climate actions, campaigners must visualise and develop these identities at different place-based communities (see Chapter 2)—and to forge solidarities among these approaches (see Chapter 5). What is key, nonetheless, is to highlight not only the threatening visions of the future but also the positive solutions to counter them (see Chapter 3).

The focus on climate solutions rather than on the debilitating pessimism of climate impacts receives strong support from the psychology and behavioural studies literature (Stoknes 2015; cf. Parag et al. 2011; cf. Smith and Leiserowitz 2013; see Chapter 3)—and remains vital in identity-building. As human beings, we are innately impatient, preferring immediate over delayed benefits (Strathman et al. 1994). Since climate actions today will largely be for the benefit of future generations, the movement must develop strategies that focus the benefits of effective climate actions into the present in developing visions of the future (cf. Chapter 3).

In framing present benefits and embedding them in the movement's visions, climate communication campaigns must adopt narratives of hope, the sense of the possibility, pride, and gratitude that come with many effective climate mitigation activities rather than placing much emphasis on dire messages, which have been proven to induce guilt, shame, and/or anxiety (Markowitz and Shariff 2012; cf. Chapter 3). Indeed, the most engaging images are not those that convey fear, but optimistic icons that make people feel they are able to do something about it (O'Neill and Nicholson-Cole 2009; Kahan and Braman 2008; Floyd et al. 2000). Marshall Ganz (2006) highlights the importance of using these types of hopeful narratives:

> Devising a credible strategy and telling a motivational story go together. Most effective campaigns have a complementary 'story' and 'plan.' How we can build from resources we have, how we can take advantage of opportunities, why the constraints will not overwhelm us, how each steps leads to the next – all of these are elements in a plausible strategy. Just as

good strategy gives individual tactics meaning by transforming them from isolated events into steps on the road to our goal, a good story gives our actions meaning by transforming us into participants in a powerful narrative… "deconstruct" an old story, on the way to learning a new one.

This means campaigners have to carefully weave credible stories involving repercussions of unabated emissions with positive discourses on the contributions of effective mitigation efforts at the local level, particulalry to health, jobs, sustainability, democracy, and empowerment, among others. SAE38 agrees: "Being positive about the benefits of change rather than highlighting the negatives [is an 'effective' campaign approach]." What is lacking in many present climate actions are equal, if not greater, emphases and efforts in underlining these positive *solutions* for deeper emissions reductions and embedding them in the construction of visions—or what can be called 'motivational frames' vis-à-vis 'diagnostic frames' (cf. Snow and Benford 1988).

Instead of overreliance on confrontational politics, the movement must also present time-bound and achievable visions that diverse individuals and groups can, at least in principle, support. Optimistic visions containing storylines from the bottom-up are preferred over pre-formed ideology filtered top-down (cf. Saunders 2008; Chandler 2004). This does not mean, however, that climate solutions must be strictly bottom-up approaches; what I contend is that climate actions have to occur across different scales. Visions must encompass these scales to process mobilisations across complex, interdependent relationships (cf. Landy 2015; cf. Chapter 5).

As an example, the overarching vision of climate actions can be advanced by campaigns and activities that focus on a storyline where rapid mitigation is achieved via changes in energy technologies (or energy transitions) together with changes in consumption behaviours. Some climate action groups are already emphasising their focus on some of these alternative solutions (e.g. see Renewables 100 Policy Institute 2007). Similar campaigns, however, must be broadened to underscore the positive visions and policies to achieve them, including in sectors other than energy, and preferably in locations where people can directly participate and contribute. This approach has support from the literature. For instance, studies have shown that this kind of mobilisation tends to be affirming rather than threatening the sense of self and basic worldviews held by the audience (cf. Kahan and Braman 2008;

Floyd et al. 2000), hence are most supported. Compared to vague and scattered alternatives, these campaigns that focus on solutions, where people can translate their concern into feasible actions, are more effective.

Motivational and translational climate mitigation is already clear among some climate action groups, especially to those advocating sustainable energy transitions (e.g. see Go 100% Renewable Energy, n.d.; Renewables 100 Policy Institute 2007) and has already been included, albeit loosely, within the general goals of the climate action movement. Underlining the promise of effective climate mitigation through sustainable energy transition among others as a unified alternative must be a top priority in crafting the movement's visions of the future. Campaigning along this line can be personalised, and therefore, can hold greater salience to a variety of audiences (cf. Chapter 1). Individuals, groups, and local communities must, therefore, understand what rapid and effective climate actions entail for them and what sort of positive activities can be designed at their own sphere of influence. Careful planning of strategies, wrapped around powerful narratives that clearly outline the pathways to and outcomes of effective climate actions (cf. Chapter 2), alongside their limitations, is equally vital.

A powerful storyline that can be adopted in the movement's vision is to highlight the technical and economic feasibility of rapid transition to renewable energy at local scales, that is, in communities, towns, and cities. Already, community-based sustainable energy systems have been spreading the narratives of hope, thus building new identities that people could value in the process. Real-world evidence for the feasibility of this type of transition has already been demonstrated in the island of Samsø in Denmark; the counties of Nordfriesland, Prignitz, and Dithmarchen in Germany; and the cities of Illinois and Ithaca, New York in USA (Cleantechnica 2013; see Go 100% Renewable Energy, n.d.), to cite a few. These alternatives have also provided multiple co-benefits beyond climate actions, including the provision of new employment opportunities, inclusive community life, and a promise of a sustainable future (Van der Schoor and Scholtens 2015; Oteman et al. 2014; Seyfang et al. 2013; Hopkins 2008). Its socio-economic benefits, which include saving money on energy bills, generating income for the community, tackling fuel poverty, improving local economy, skills development, and local job creation (Seyfang et al. 2013), have also helped boost the motivational frame of this narrative. For some, underlining the economic promise of climate actions, such as these, is more important than anything else. As SAE24 suggests: "[it] seems [that] if you can get it [the campaign] into dollar terms, people respond."

Empirical data tends to support the contention that generating income for communities through these activities is, indeed, a powerful incentive for mobilisation. A UK study on community energy suggests that the aspiration to generate local income, along with ensuring community sustainability, has emerged as a far more important driver than environmental motivations such as the desire to reduce the community's carbon footprint (Bomberg and McEwen 2012). Campaigners could bank on this by underlining how these local climate actions provide simple, yet feasible, straightforward, output-driven, measurable, and verifiable options clear enough for laypersons to understand. It also can make people feel they can do something about the climate challenge (cf. Chapter 1).

In these localised, action-oriented, yet vision-based, campaigns, people can be shown how small- and medium-scale renewable energy projects can be realised in their own local spaces. A range of proven renewable energy supply technologies and systems to meet their electricity demand is already available with their cost fast declining. These options may also include energy efficiency activities and interventions to manage their demand. Systems that can be tapped through this scale include micro-scale biogas, solar PV, solar water heating, or community-owned on-shore wind farms. In case these options are not applicable, people also can pool their money for investment on solar farms, offshore wind, wave or tidal power, etc. Energy efficiency, meanwhile, can be met by neighbours doing a bulk buy of their energy, as well as in building retrofits in residential houses and buildings, community precincts, local government offices, and other public buildings such as community libraries and centres. Local public and active transport infrastructure can also be designed such that they are powered by locally generated sustainable energy sources. If replicated and scaled, community-based energy for rapid mitigation holds a highly promising approach to effective climate mitigation.

Local people can be best mobilised to do local climate actions if they are provided with immediate benefits (cf. Chapters 1, 2, and 3). Community energy solutions are most appealing in that it allows independence from electric utilities, protection of local environments, new sources of income, and increasing social cohesion (Seyfang et al. 2013;

Yagatich et al. 2018)—benefits that are highly attractive to local people. Additionally, micro-level transitions in cities (North and Longhurst 2013; Evans 2011) and even in rural areas (Lovell et al. 2018; Delina 2018) also provide spatial contexts in which experimentations and learning with new economies can be articulated and, hence, climate actions can be mobilised. With community-based actors experimenting with innovations for transition to low-carbon societies, large-scale transitions can then be accomplished virally as thousands of other individuals and groups follow suit. Of course, not all these options can be successful; in fact, failures are more common than successes. Nevertheless, these experiences offer the movement an opportunity to reflect on their campaigns—successful and not—and, thus, strengthen future efforts. While this kind of future is ideal, a "blooming of a thousand flowers" has already permeated many climate actions in many local governments where alternative pathways have been occurring through a multiple, yet structurally webbed, local energy transitions (cf. Chapter 5). The Transition Towns movement serves an example demonstrating a number of feasible small- and medium-scale, community-based sustainable energy solutions through renewable energy, energy efficiency, and energy conservation activities (Hopkins 2008).

The climate action movement must also incorporate climate actions beyond community energy projects in their campaign agenda to account for various nuances and heterogeneity of actors in the movement. Intersectionality of overlapping motivations increases the opportunity for participation (Fisher et al. 2017; Adam 2017; cf. Chapter 5). Indeed, not everyone will be presented the opportunity to directly participate in community energy transitions. Fossil fuel divestment, a strategy first suggested by 350.org, offers one possibility for engagement. Divestment has, despite differing motivations of its promoters, already, propelled itself into a successful global climate action campaign (see Chapter 2). Ensuring these many prefigurative climate actions are as diverse as possible, sustained for their viability and productivity, and replicated as rapidly as possible remains a paramount mobilisation agenda, and, therefore, has to be embedded in the movement's vision documents.

Climate action groups can focus their campaigns on pushing more local governments and councils to adopt pathways that lead to climate actions. The focus on local governments is paramount since support for climate actions at this level of government is relatively higher than at the national and federal levels. Rapid emission reductions at this level also are promising especially since cities account for approximately 70% of GHG emissions (United Nations Human Settlements Programme 2011). According to a 2014 report by C40, the United Nations Secretary General's Special Envoy for Cities and Climate Change, and the Stockholm Environment Institute (2014), cities could dramatically reduce their emissions by adopting policies that are within their authority. The report estimates that by 2030, cities could achieve 10% of the global emissions reductions to close the gap between our present trajectory and what is required to stay below 2 °C in warming, and by 2050 that figure could increase to 15%.

Stories of successful climate actions in cities abound. In 2010, the city council of Eugene, Oregon in USA passed an ordinance to achieve steep reductions in energy consumption and GHG emissions in both private and public sectors by legislating the city's Community Climate and Energy Action Plan (City of Eugene 2010). The City Government of New York (2014) also announced the city's commitment to reduce its emissions 80% below 1990 levels by 2050 by rolling out a comprehensive plan to improve energy efficiency in buildings. Since then, many cities have followed suit and have even linked their many actions. Even earlier, the world's largest cities formed the *C40 Cities Climate Leadership Group* (C40) in 2005 to coordinate their efforts and learn from one another. Approximately 69 affiliated cities from more than 40 countries, accounting for a twelfth of the global population, now comprise C40 (C40 2015). In September 2014, during the UN Climate Summit, the mayors of Los Angeles, Houston, and Philadelphia in USA announced a *Mayors' National Climate Action Agenda*[3] to set targets in their cities for emission reductions, and to create or update climate action plans (Barboza 2014). Also during the Summit, the formation of a *Compact of Mayors*[4] was announced to address climate change at the city level by bringing in information on best practices, fostering competition, and modelling the good behaviour for national governments to adopt (C40, United Cities and Local Governments, and International Council for Local Governments, n.d.).

---

[3] See full text of the initiative at City of Houston (n.d.).

[4] See full text of this agreement at C40, United Cities and Local Governments, and International Council for Local Governments (n.d.).

Visions of sustainable energy transitions that first sprung from local-based transitions provide one of the many optimistic demonstrations of effective climate actions. These images, which show the possibility of effective climate actions through sustainable energy transitions—from community to community, town to town, city to city, and eventually from country to country—provide the climate action movement an enduring demonstration of hope that helps trigger a new identity hence stimulating greater mobilisation. The climate action movement should continue using these images to highlight and demonstrate the availability of alternatives to the fossil fuel regime.

In closing, a vision of a durable and sustainable climate safe future is vital in mobilisation. This chapter describes key points surrounding processes of climate action visioning: building a unified regime alternative; emphasising positive solutions over threatening visions; and highlighting optimistic storylines of solutions. This chapter highlights more crystallised and invigorated local climate actions to spur a new identity by which people can see themselves supporting. Visions of optimistic futures—conveyed through climate actions that focus on proximate, positive solutions—offer a framework for such kind of mobilisation. Campaigners, however, are cautioned that simple reframing of climate solutions is still unlikely to increase climate action engagement. The multiple ways by which publics are mobilised for climate actions remain heterogeneous at best. While these triggers for climate actions may be plural and are structurally disconnected from each other, they can still be bridged, amplified, and extended to deliver large-scale mobilisation. The next chapter covers some of these dynamics of extension.

References

Adam, E. M. (2017). Intersectional coalitions: The paradoxes of rights-based movement building in LBTQ and immigrant communities. *Law & Society Review, 51,* 132–167.

Barboza, T. (2014, September 22). L.A., Houston, Philadelphia mayors vow more action on climate change. *Los Angeles Times.* http://lat.ms/1odyfOn.

Bomberg, E., & McEwen, N. (2012). Mobilizing community energy. *Energy Policy, 51,* 435–444.

C40 Cities Climate Leadership Group (C40). (2015). *C40 Cities.* http://bit.ly/1ir0TJ5.

C40, UN Secretary General's Special Envoy for Cities and Climate Change, & Stockholm Environment Institute. (2014, September). *Advancing Climate Ambition: Cities as Partners in Global Climate Action*. http://bit.ly/1yGBKXC.

Chandler, D. (2004). Building global civil society 'from below'? *Millennium: Journal of International Studies, 33,* 313–339.

City Government of New York. (2014). *One City, Built to Last: Transforming New York City's Buildings for a Low-Carbon Future*. http://on.nyc.gov/1pJQQRB.

City of Eugene. (2010). *A Community Climate and Energy Action Plan for Eugene*. http://bit.ly/1vEpT6r.

Cleantechnica. (2013). *Germany: 100% Renewable Energy and Beyond*. RenewEconomy. http://bit.ly/12e7iT6.

Delina, L. (2018). Can energy democracy thrive in a non-democracy? *Frontiers in Environmental Science, 6,* 5.

Delina, L., Diesendorf, M., & Merson, J. (2014). Strengthening the climate action movement: Strategies from histories. *Carbon Management, 5,* 397–409.

Della Porta, D. (2005). Multiple belongings, tolerant identities, and the construction of 'another politics': Between the European social forum and the local social fora. In D. Della Porta & S. Tarrow (Eds.), *Transnational Protest and Global Activism*. Lanham, MD: Rowman & Littlefield.

Della Porta, D., & Diani, M. (2006). *Social Movements: An Introduction* (2nd ed.). Malden, MA: Blackwell.

Diani, M. (1992). The concept of social movement. *The Sociological Review, 401,* 1–25.

Diani, M. (1995). *Green Networks*. Edinburgh, UK: Edinburgh University Press.

Evans, J. P. (2011). Resilience, ecology and adaptation in the experimental city. *Transactions of the Institute of British Geographers, 36,* 360–376.

Fisher, D. R., Dow, D. M., & Ray, R. (2017). Intersectionality takes it to the streets: Mobilizing across diverse interests for the Women's March. *Science Advances, 3,* eaao1390.

Floyd, D. L., Prentice-Dunn, S., & Rogers, R. W. (2000). A meta-analysis of research on protection motivation theory. *Journal of Applied Social Psychology, 30,* 407–429.

Ganz, M. (2006b). Mobilizing power: Analysis, strategy, deliberation. In *Organizing Course Notes*. Cambridge, MA: Harvard Kennedy School. http://bit.ly/1xcbrGW.

Hopkins, R. (2008). *The Transition Handbook: From Oil Dependency to Local Resilience*. Totnes, Devon, UK: Green Books.

Kahan, D. M., & Braman, D. (2008). The self-defensive cognition of self-defense. *American Criminal Law Review, 45,* 1–65.

Landy, D. (2015). Bringing the outside field interaction and transformation from below in political struggles. *Social Movement Studies, 14,* 255–269.

Lorenzoni, I., Nicholson-Cole, S., & Whitmarsh, L. (2007). Barriers perceived to engaging with climate change among the UK public and their policy implications. *Global Environmental Change, 17,* 445–459.

Lovell, H., Hann, V., & Watson, P. (2018). Rural laboratories and experiments at the fringes: A case study of a smart grid on Bruny Island, Australia. *Energy Research & Social Science, 36,* 146–155.

Markowitz, E. M., & Shariff, A. F. (2012). Climate change and moral judgment. *Nature Climate Change, 2,* 243–247.

Melucci, A. (1989). *Nomads of the Present.* Philadelphia: Temple University Press.

Melucci, A. (1996). *Challenging Codes.* Cambridge, UK: Cambridge University Press.

Moser, S. C. (2009). Costly knowledge—Unaffordable denial: The politics of public understanding and engagement on climate change. In M. T. Boykoff (Ed.), *The Politics of Climate Change: A Survey* (pp. 155–181). Oxford, UK: Routledge.

North, P. (2011). The politics of climate activism in the UK: A social movement analysis. *Environment and Planning A, 43,* 1581–1598.

North, P., & Longhurst, N. (2013). Grassroots localization? The scalar potential of and limits of the 'transition' approach to climate change and resource constraint. *Urban Studies, 50,* 1423–1438.

O'Neill, S. J., & Nicholson-Cole, S. (2009). "Fear won't do it": Promoting positive engagement with climate change through visual and iconic representations. *Science Communication, 30,* 355–379.

Oteman, M., Wiering, M., & Helderman, J.-K. (2014). The institutional space of community initiatives for renewable energy: A comparative case study of the Netherlands, Germany and Denmark. *Energy Sustainability and Society, 4,* 11.

Parag, Y., Capstick, S., & Poortinga, W. (2011). Policy attribute framing: A comparison between three policy instruments for personal emissions reduction. *Journal of Policy Analysis and Management, 30,* 889–905.

Renewables 100 Policy Institute. (2007). *What We Are.* http://bit.ly/15AqFL8.

Saunders, C. (2008). Double-edged swords? Collective identity and solidarity in the environmental movement. *The British Journal of Sociology, 59,* 227–253.

Schock, K. (2005). *Unarmed Insurrections: People Power Movement in Nondemocracies.* Minneapolis, MN: University of Minnesota Press.

Seyfang, G., Park, J. J., & Smith, A. (2013). A thousand flowers blooming? An *Examination of Community Energy in the UK, Energy Policy, 61,* 977–989.

Smith, N., & Leiserowitz, A. (2013). The role of emotion in global warming support and opposition. *Risk Analysis, 34,* 937–948.

Snow, D. A. (2013). Framing and social movements. In D. A. Snow, D. della Porta, B. Klandermans, & D. McAdam (Eds.), *The Wiley-Blackwell Encyclopedia of Social and Political Movements*. Malden, MA: Blackwell.

Snow, D. A., & Benford, R. D. (1988). Ideology, frame resonance, and participant mobilization. *International Social Movement Research, 1,* 197–217.

Snow, D. A., Rochford, B., Worden, S. K., & Benford, R. D. (1986). Frame alignment processes, micromobilization, and movement participation. *American Sociological Review, 51,* 464–481.

Stoknes, P. E. (2015). *What We Think About When We Try Not to Think About Global Warming.* White River Junction, VT: Chelsea.

Strathman, A., Gleicher, F., Boninger, D. S., & Edwards, C. S. (1994). The consideration of future consequences: Weighing immediate and distant outcomes of behaviour. *Journal of Personality and Social Psychology, 66,* 742–752.

Tilly, C. (1995). *Popular Contention in Great Britain, 1754–1834.* Cambridge, MA: Harvard University Press.

Tilly, C. (2002). *Stories, Identities and Political Change.* New York: Rowman & Littlefield.

Tilly, C. (2008). *Contentious Performances.* Cambridge, UK: Cambridge University Press.

United Nations Human Settlements Programme (UN-Habitat). (2011). *Hot Cities: Battle-Ground for Climate Change.* Nairobi, Kenya: UN-Habitat.

Van der Schoor, T., & Scholtens, B. (2015). Power to the people: Local community initiatives and the transition to sustainable energy. *Renewable and Sustainable Energy Reviews, 43,* 666–675.

Yagatich, W., Galli Robertson, A. M., & Fisher, D. R. (2018). How local environmental stewardship diversifies democracy. *Local Environment, 23,* 431–447.

CHAPTER 5

Webbing

Abstract Diversity of participation is achieved when the campaign involves as many participants as possible and encompasses genders, socio-economic classes, religions, beliefs, etc. Equally essential for success, however, is how these heterogeneous actions can be linked together. This chapter calls for webbing and, so, asks how could networks be established in ways that encourage, engage, and empower campaigners and groups instead of weakening and dispiriting them. It describes how non-hierarchical, centerless, and submerged forms of resistance can be triggered, sustained, and scaled. As these dynamics pan out, this chapter further asks what contemporary social action campaigns tell of group independence and dependence, and what does this mean in terms of webbing divergent, multi-level, and transnational, yet rhisomatic, climate actions.

Keywords Networking · Webbing · Flat organisation · Linkages Coalitions · Networks

I think the greatest benefit of being part of alliances is simply the sense that we're not operating in a vacuum - that there are other people and groups out there who share our concerns, ideals, etc.
  —Respondent from an Australia-based group formed in 2007 and engages with the community and people in positions of influence to encourage them to take stronger climate actions.

© The Author(s) 2019   59
L. L. Delina, *Climate Actions*,
https://doi.org/10.1007/978-3-319-91884-6_5

Numbers are essential in social movements (Tilly 1995, 2002, 2008)—and can only be achieved once multiple climate actions are somehow linked in solidarity networks. This process can be called 'webbing.' Interconnecting diversity of participation and tactical strategies is important since climate change transcends race, gender, or region, and extends its scope to the whole of humanity, including the unborn and even non-human species (cf. Tarrow 2005; Della Porta and Tarrow 2005). The multiplicity of possible climate actions that can be spurred across many spaces is another consideration. A climate action movement, hence, can be best described as a constellation of many heterogeneous actors and groups with differing emphases on the nature of the challenge and unique capacities, but somehow coordinated and eventually linked (cf. Saunders 2013). The present climate action movement, indeed, has been constructed as a web of non-government organisations; anarchist and autonomist grass-roots groups; traditional labour groups; left political parties; emancipatory movements around questions of identity; religious and faith-based groups; professional organisations; and indigenous peoples, among many others. This illustrates solidarities across diversity (Routledge 2012). Webbing manifests in convergences, affiliations, and networks, and evidences a show of force through trans-local or trans-national solidarities (Nunes 2009). Such processes of webbing, however, have to occur despite the absence of formal institutional arrangements, a hierarchy, and central authority (Della Porta and Diani 2006; Andrews et al. 2010).

Webbing provides an umbrella to focus different forms of climate actions that individuals and groups carry out at different times under different organisations using different tactics targeted on different audiences (Schock 2005). Through a networked approach, climate actions can be pursued within and across individual industry sectors such as energy, transport, forestry, and agriculture, or holistically through promoting green technologies across the economy, or even going beyond technologies to stop the growth in economic activity and population which are, at one conceptual level, two of the three drivers of environmental destruction (Dietz and O'Neill 2013; Daly and Farley 2004). A heterogeneous, yet webbed, response is also important to address the free-rider problem (Polleta and Jasper 2001).

The more diverse the participants are—in terms of gender, age, religion, ethnicity, ideology, profession, socio-economic status, etc.—the better the chance of success. This diversity also makes it more difficult for adversaries to isolate climate action adherents (Delina et al. 2014). Most respondents agree that the diversity of their participants is vital ($N = 42$, median $= 4$, Interquartile range (IQR) $= 1$[1]). Twenty-five respondents or 60% reveal that having diverse participants has led to 'effective' campaigns, while 13 or 31% state a diverse constituency leads to 'very effective' campaigns. SAE34[2] highlights the significance of diversity in "getting people from different constituencies at the table, making sure all voices are heard, and all parties are involved in creating plans of action, including implementation accountability." Diversity of participation also increases the likelihood of tactical diversity, since different groups are familiar with different forms of campaigns and bring their own capacities to their respective movements (Murphy 2005; Chenoweth and Stephan 2011). SAE36, for instance, notes: "The fossil fuel divestment movement seems to have more legitimacy and credibility due to the diversity of campaigns, organisers, and supporters involved."

Webbing opens the movement to a variety of people who have a variety of reasons for participating. This is key since motivations, interests, values, hopes, priorities, and dreams differ. Some join because they see their friends and families in those groups (Fisher and McInerney 2012; cf. Chapter 1). Others are motivated by altruism, social solidarity, love of nature, concern for future generations and community values, etc. (Seyfang et al. 2013; Hall and Taplin 2008). Others are attracted to the movement because of the economic opportunities and jobs that come along with the campaigns (Seyfang et al. 2013; cf. Chapter 4). This heterogeneity suggests campaigners cannot rely on one particular rationale to attract participation (cf. Chapter 1). Climate change must be connected to relevant issues that people care about to provide wide-ranging opportunities for heterogeneous audiences to take action (cf. Chapter 2). Webbing allows campaigns to be structured so people can put in as many or as few resources as they are able, while still making a contribution.

---

[1] See Appendix B for notes on statistical treatment and reporting.
[2] See Table C.1 for the description of the social action groups.

Webbing is achieved through interconnectedness in alliances, coalitions, or networks (Featherstone 2005; Schock 2005; Tarrow 1998; Tilly 1995) to generate productive connections (Featherstone 2008). Citing examples from their case studies of large-scale social mobilisations in history, Delina et al. (2014) argue how the networks of local-based groups worked with and supported each other to achieve their bigger campaign objectives. In their Philippine case study, for example, they highlighted how the Roman Catholic Church mobilised groups of Filipino peasants, farmers, and fisherfolk years before the anti-Marcos movement had achieved the generally peaceful People Power Revolution in 1986. By contrast, their case study from Burma showed how the neglect of webbing opportunities across the Burmese society failed to deliver the Burmese ideals of democracy in late 1980s.

Network-based climate actions have already produced many outward-focused climate activism. On 21 September 2014, the power of the networked approach was evidenced in one of the largest gatherings of the climate action movement. 350.org and a number of other organisations coordinated the event dubbed the People's Climate March (Foderaro 2014). Outward-oriented climate activism such as this must be sustained and strengthened by encouraging more groups to participate in network-wide activities, and by organisers in the movement designing more large-scale climate actions of this type. It indeed remains paramount and relevant that campaigners continue to expose the tactics of vested interests and to conduct nonviolent resistance to greenhouse pollution, through interconnected climate actions.

Our survey respondents underline the importance of webbing where thirty-five respondents or 75% state their groups are affiliated with networks, coalitions, or alliances. Most respondents, including some of those who do not have any affiliation (twelve or 26%), strongly agree with the idea that affiliations could result in more 'effective' campaigns ($N = 46$; median $= 4$; IQR $= 1$). SAE27 notes "building strong commitment within social networks and then building support network-by-network is vital." Thirteen respondents or 28% state campaigns could be weakened by non-alliance with like-minded groups. Failure to organise joint campaigns and protest actions with other groups, twenty-four respondents or 51% state, could result in 'ineffective' campaigns. Table 5.1 shows some of the affiliations of the survey respondents.

While network-organised, outward-focused climate actions already tend to be common, forging many networks of individuals, groups, and

Table 5.1 Affiliations of responding social action groups to networks, coalitions, or alliances

Group	Affiliation
SAE01	Alliance for the Common Good; Dawnland Defense; Natural Resources Council of Maine; Natural Resources Defense Council (NRDC); Protect South Portland; Seeds for Justice; Interfaith Power and Light; Sierra Club-Maine Chapter; Environment Maine; Stop the East-West Corridor
SAE03	Climate Action Network Australia (CANA); 350.org; Leard Forest Alliance
SAE04	Massachusetts Pipeline Awareness Network; Teaming with Wildlife; Massachusetts Rivers Alliance; Massachusetts Land Trust Coalition
SAE05	Sierra Student Coalition
SAE06	U.S. Climate and Health Alliance (CAHA); Chicago Clean Power Coalition; RE-AMP; U.S. Climate Action Network (USCAN)
SAE07	Heal; Green Budget; European Environmental Bureau
SAE08	CANA
SAE09	Climate Action Network (CAN) International; CAN South Asia; Climate Change Network Nepal; NGO Network on Climate Change; Agriculture for Food Alliance
SAE10	Northwatch; Ontario Environmental Network; Green Communities
SAE11	CANA; Australian Youth Climate Coalition (AYCC); Citizens Own Renewable Energy Network (CORENA); The Goulburn Group
SAE12	CANA; Protect Sydney's Water; Our Land Our Water Our Future
SAE13	CANA
SAE17	Dogwood Alliance; NRDC
SAE19	Australian Network of Environmental Defenders Offices; Tasmanian Coastal Alliance; Community Legal Centres Tasmania
SAE21	Canadian CAN; Toronto CAN
SAE22	CANA; Victorian CAN
SAE23	CAN Europe; Butterfly Conservation Turkey; İkli Ağı (Climate Network)
SAE25	Canadian CAN; Global Call for Climate Action; Centre for Social Innovation
SAE26	Pennsylvanians Against Fracking; Protecting Our Children; Stop the Frack Attack; Americans Against Fracking
SAE27	Transition Decade Alliance
SAE33	· CANA; North East Regional Sustainability Alliance; Environment Victoria
SAE34	USCAN; Federal Forest Carbon Coalition; Creation Care Coalition
SAE36	Fossil Free; 350.org; Students Divestment Network; Ivy Divest
SAE38	CANA; Environment Tasmania; Sustainable Living Tasmania
SAE39	Divest/Invest; California Student Sustainability Coalition
SAE40	None reported
SAE41	Network of climate emergency groups; CANA

(continued)

Table 5.1   (continued)

Group	Affiliation
SAE42	National Organization of Physicians for Social Responsibility; Coalition efforts with Sierra Club, Sustainable Tucson, Tucson 2030 District, Center for Biological Diversity, Earthjustice, Tucson Clean and Beautiful, and others
SAE43	Coalition for Community Energy Australia
SAE44	Ontario Environmental Network; Crane Institute
SAE45	CAN Europe; Butterfly Conservation Turkey; İkli Ağı (Climate Network)
SAJ01	The Refugee Council; The Australian Coalition to End Child Detention; Australian Catholic Religious Against Trafficking in Humans (ACRATH)
SAO01	Global CAHA; Canadian Partnership for Children's Health and Environment; International Society of Doctors for the Environment; Canadian CAN
SAO02	Media Consortium
SAO03	Global CAHA; CANA; Health Care Without Harm
SAO04	Climate and Energy Literacy Network
SAO05	Sierra Club; Clean Air Alliance; Clean Air Action; Tucson CAN; Citizens' Climate Lobby
SAO06	Australian Religious Response to Climate Change; Pacific Partnership Calling; Faith Ecology Network; ACRATH; Catholic Religious Australia
SAO07	Integral; National Council for Voluntary Organisations; Micah Network; The Climate Coalition; CAN International; Publish What You Pay; Network of International Development Organisation in Scotland
SAO08	CANA; Global Climate and Health Alliance; Healthy Energy Initiative
SAO09	American Association for the Advancement of Science; National Science Teachers Association; also frequently works with ad hoc coalitions

communities that focus on prefigurative activism—climate actions that demonstrate options and alternatives to the status quo—has yet to be accomplished at a comparable scale. On that regards, David Hess (2018) suggests building strong energy transition multi-coalitions. There already exist models for achieving this end, including examples from community energy groups that are collectively working and coordinating their activities (e.g. Bristol Energy Network and the Community Energy Scotland Network in the UK (also see Parag and Janda 2014), various energy networks in The Netherlands (Van der Schoor et al. 2016), and the 100% Nachhaltige Energie Regionen in Germany (Beveridge and Kern 2013), among others. More affiliations such as these are required to diffuse similar prefigurative climate actions widely. Similarly needed are interconnections across climate actions that address overconsumption.

Networked climate actions must be integrated for them to successfully diffuse nationally and transnationally. To do this, Hess (2018) suggests examining the conditions under which integration can appear (e.g. the importance of a bridge organisation to broker a coalition, the use of the frame of energy democracy, etc.). In addition, an integrated approach to coordination and communication within and among climate action networks is imperative. This would ideally require setting national and international hubs. Information flow of the diverse actions by climate action groups can be facilitated within and between these hubs. The hubs also can function as information clearinghouses where relevant research results and examples of effective practices can be compiled so they can be translated to actual climate actions in other locations.

To retain this success in future climate action mobilisations, a sustained, dynamic, and cohesive public engagement remains necessary, highlighting the essential role of webbed climate action messages widely and effectively in plural societies. Work must extend beyond the movement's original constituency to include issues thought to be relevant to bystanders and other potential climate action adherents (Snow et al. 1986).

The processes of webbing have received mixed sentiments and views, and webbing effects have been varied to our respondents. Most respondents indicate such affiliations make them feel their group can now make greater impact since they have become part of or affiliated with a larger group (2014 survey: $N = 35$, median $= 3$, IQR $= 1$). Seventeen respondents or 49% agree with this statement, while 16 or 46% are in 'strong' agreement. Along the same vein, membership with 'umbrella' organisations make most respondents feel they have become a stronger force ($N = 35$, median $= 3$, IQR $= 1$) as a critical mass is formed around certain goals (SAE03). Twenty-one respondents or 60% agree with this statement, while 11 or 31% are in 'strong' agreement.

Among the most important benefits of affiliations is to have a shared notion that would potentially create a common ground (see Chapter 4). This enables varied themes to be interconnected and for different groups and actors from various struggles and social contexts to join in the struggle (Della Porta et al. 2006). SAE12 states: "Being part of a larger alliance has … allowed [us] to be part of a bigger campaign, boosting impact." Converging around a common struggle (Cumbers et al. 2008) reduces campaigners' feelings of isolation. SAE31 shares this sentiment noting how webbing "makes us feel we are not crying in the wilderness and are not mad." SAE08 echoes the same: "Good to feel [that] you are not alone and that there are other passionate people out there running their own campaigns. Together we make much more noise." The benefits are especially present in specific campaign strategies such as, for example, in SAE39's fossil fuel divestment campaign:

> Divestment necessitates a critical mass to be an effective means of social (or especially financial) stigma, so being intimately involved in coalition with other students, frontlines, and financial campaigns against fossil fuel extraction and consumption is [sic] greatly beneficial. Divestment also acts as an organizing tool more than [an] end-goal; it is a way to build these relationships and coalitions to ensure climate justice and equitable distribution of sacrifice as our climate changes.

Another benefit of affiliation is the opportunity to share the burdens of campaigning (SAE22). SAE34 notes: "The problem is too big to face alone. Coalitions are really the only way." SAE27 explains:

> The crucial value of an alliance is that it can create a larger virtual organisation but has the value that the sub-units can specialise their work (in the knowledge that other crucial functions are being carried out by other members of the alliance) and that each unit can be self-managing thus avoiding managerial overload and excessive concentration of power in the hands of a few.

Access to resources also recurs as one of the positive effects of interconnections among the responses. Most respondents agree their group can now combine their own resources with other groups' ($N = 35$; median $= 3$ (agree); IQR $= 1$). Some groups have even found new streams of resources and strategies because of these affiliations (SAE10, SAE12, and SAE25), including sources of funding (SAE11) and opportunities for sharing, learning, and collaboration (SAE21). As SAE22 puts it: "No re-inventing - rather hop on board work done by others and share the burden."

Despite these benefits, however, some respondents report disadvantages. For instance, twenty respondents (57%) state webbing resulted in some feelings they are somehow losing part of their 'independence.' SAE06, for instance, shares an example showing how challenging it could be for 'umbrella' organisations to make one group's specific concern shared across network members: "It's sometimes hard to get other groups to take the public health effects of climate change seriously. Lots of lip service around this issue—our core issue—not a lot of action." Retaining a sense of autonomy, while valuing collaboration, is indeed important (Rosewarne et al. 2014: 84–86). For instance, SAE01 shares:

> We struggle with the lack of shared values. Yes, we can unite on a specific issue but often our environmental colleagues want us to model corporate hierarchy or privilege in how we deal with aspects of our work together. The members on the frontlines of coalitions have the most difficult time with this and are often tempted to pull out of coalition. Usually feelings calm down and we feel the advantages outweigh the disadvantages, but from time to time we have to question who we are and what we stand for. We have pulled out of some group decisions and limited our coalition involvement in these instances.

SAE04 raises an observation as to who gets the credit at the end of a successful campaign:

> We need to carefully balance each group's claim of credit for successes in our coalition - make sure everyone gets credit. This is the hardest part of the balance. Keeping our board members happy that we are making a huge difference without them pushing or have us claim all the credit, or getting mad when coalition members claim more credit (or can raise more money).

Sometimes, the nature and focus of work that defines a particular social action group also limits the advantages of affiliation. While stressing the paramount implication of webbing, SAJ01, for instance, mentions the challenge of being a faith-based group:

> It is challenging for religious groups to join civil society coalitions - but it is also essential in terms of resources and reach. The difficulty comes from retaining our ethos and ensuring we don't alienate those who support us because we are a religious organization... while still ensuring that we are relevant, accessible and heard by non-religious members within the coalition and the coalition's supporter base.

Managing affiliations is another challenge as SAE19 emphasises:

> The additional logistics involved in working in a network can be a bit daunting or time consuming - it has sometimes felt like the extra effort was not worth the return.

Perhaps one important caveat to note in understanding the process of webbing is the resultant condition brought about by coalition-building to the larger movement. As some of the respondents emphasise just above, there could be the danger of a divided movement as a 'we-them' dichotomy is constructed between organisations within the same movement during processes of coalition-building (Saunders 2008). The concept of multistakeholderism illustrates this. Multi-stakeholder engagement is often tagged as necessary to build more inclusive, participatory, multi-scale, and heterogeneous climate actions. However, multistakeholderism does not only focus on plurality per se but also on the differences on identities, interests, roles, and responsibilities. Quite often, the power imbalances arising from these pursuits are ignored within a social movement. For this reason, organisations may not coalesce with each other that easily. In the UK, Saunders (2009) found that groups with moderate action repertoire and constructive relationship with government tend not to cooperate with groups having radical action repertoire and negative relations with the state. The relationships when foxes and chickens share the same coop, therefore, remain a question that will beleaguer climate action campaigners for some time.

Within the broad climate action movement itself, divisions by typologies are already present (see for e.g. Hess 2018). For instance, some groups value linkages with politicians while others see this type of partnership as unacceptable. Most respondents indicated a strong agreement with the idea that if their group is seen as an ally of a political party, their campaigns tended to be 'ineffective,' meaning lesser public support and participation ($N = 32$, median $= 3$, IQR $= 1$). Fifteen respondents or 47% declare allying with a political party as a 'very ineffective' approach, while 14 or 44% see it as 'ineffective.' Some, however, think otherwise. While a number of respondents look askance at political party affiliation, twenty-nine respondents or 62% identified that the absence of politician's support has been one of the many barriers for 'effective' campaigns. To that end, emphases towards creating communication channels through "lobbying of politicians at all levels—local government, state and federal" (SAE22, SAO03, and SAO07); "meeting with policy-makers (and) elected representatives" (SAO01); and "finding a leader in politics to champion our cause" (SAE10) are paramount.

Webbing with the business sector is another contentious issue. The push for climate legislation in 2008 in USA, where webbing with businesses through the United States Climate Action Partnership (USCAP), provides an example. USCAP was an alliance among leading US-based environmental movements and businesses formed to push a cap-and-trade system in the US Congress (USCAP 2009a). The alliance includes the prominent Environmental Defense Fund, World Resources Institute, and National Resources Defense Council in its membership roster, and those in the fossil fuel business such as Rio Tinto, Duke Energy, and Shell (USCAP 2009b). Despite the seemingly strong interconnections across multiple stakeholders, USCAP failed to advance a cap-and-trade bill in the US Congress.

In her analysis of the failed alliance, Theda Skocpol (2013) argues that if these highly respected environmental groups concentrated instead on soliciting and mobilising grass-roots support, rather than allying with and making an 'insider deal' with the business sector, results could have been different. Skocpol (2013) based her assertion on contrasting the efforts by USCAP and those who lobbied for a universal health care, later known as Obamacare. Skocpol (2013) argues that the inclusion of a strategy linking grass-roots players with those at the national and state levels through a networked approach to promote the idea of a public option led to the success of Obamacare. This proposal would make cheaper health insurance available for uninsured Americans who cannot afford private health insurance by creating a government-run health insurance agency to compete with privately held health insurance companies. Although public option was not included in the final version of the bill, the idea was considered to have successfully provided broader public support for the bill (Plumer 2013).

Journalist and activist Naomi Klein shares Skocpol's views. In an interview, Klein remarked:

> Their [pertaining to USCAP] so-called win-win strategy has lost. That was the idea behind cap-and-trade. And it was a disastrously losing strategy. The green groups are not nearly as clever as they believe themselves to be. They got played on a spectacular scale. Many of their partners had one foot in [USCAP] and the other in the U.S. Chamber of Commerce. They were hedging their bets. And when it looked like they could get away with no legislation, they dumped [USCAP] completely. (see Mark 2013; cf. Klein 2014: 226–228)

Acknowledging Skocpol's thesis, 350.org's Bill McKibben (2013) highlights how climate action networks such as 350.org have been working to revive a networked and locally organised grass-roots movement for climate actions. These climate action networks are already far-reaching and many are being forged among other groups. The Climate Action Network (CAN) International and its branches in Australia, Canada, Europe, and elsewhere are some examples.

In closing, linkages and interconnections matter. Webbing provides a platform for diversity to thrive. It allows convergence of the varied values, interests, perspectives, and capabilities innate in a social movement. The multiplicity of the challenges, approaches, scale, scope, timing, and solutions means that the climate action movement has to recognise and include varied values, interests, perspectives, worldviews, capacities, and capabilities in strategising for climate actions (cf. Saunders 2009). Responding to this heterogeneity is important in strengthening the climate action movement since people from multiple levels and scales, locations, with varying interests, priorities, aspirations, hopes, and dreams, may join climate actions for different reasons. The climate action movement, thus, cannot rely on one particular rationale to attract participation and enable coalition-building. Webbing ideally involves campaigns to be interconnected so that individual and group activities, which may seem "too small to make a difference," can multiply and thus spread their benefits quickly.

References

Andrews, K. T., Ganz, M., Baggetta, M., Han, H., & Lim, C. (2010). Leadership, membership, and voice: Civic association that work. *American Journal of Sociology, 115,* 1191–1242.

Beveridge, R., & Kern, K. (2013). Energiewende in Germany: Background, developments and future challenges. *Renewable Energy Law Policy Review, 1,* 3.

Chenoweth, E., & Stephan, M. J. (2011). *Why Civil Resistance Works: The Strategic Logic of Nonviolent Conflict.* New York: Columbia University Press.

Cumbers, A., Routledge, P., & Nativel, C. (2008). The entangled geographies of global justice networks. *Progress in Human Geography, 32,* 183–201.

Daly, H., & Farley, J. (2004). *Ecological Economics: Principles and Applications.* Washington, DC: Island Press.

Della Porta, D., & Tarrow, S. (Eds.). (2005). *Transnational protest and global activism.* Lanham, MD: Rowman & Littlefield.

Delina, L., Diesendorf, M., & Merson, J. (2014). Strengthening the climate action movement strategies from histories. *Carbon Management, 5,* 397–409.

Della Porta, D., & Diani, M. (2006). *Social Movements: An Introduction* (2nd ed.). Malden, MA: Blackwell.

Della Porta, D., Andretta, M., Mosca, L., & Reiter, H. (2006). *Globalization from Below*. London, UK: University of Minnesota Press.

Dietz, R., & O'Neill, D. (2013). *Enough Is Enough: Building a Sustainable Economy in a World of Finite Resources*. San Francisco, CA: Berrett-Koehler.

Featherstone, D. (2005). Towards the relational construction of militant particularisms: On why the geographies of past struggles matter for resistance to neoliberal globalisation. *Antipode, 37*, 250–271.

Featherstone, D. (2008). *Resistance, Space and Political Identities: The Making of Counter-Global Networks*. Oxford, UK: Wiley-Blackwell.

Fisher, D. R., & McInerney, P.-B. (2012). The limits of networks in social movement retention: On canvassers and their careers. *Mobilization: An International Journal, 17*, 109–128.

Foderaro, L. W. (2014, September 21). Taking a call for climate change to the streets. *The New York Times*. http://nyti.ms/1qkVZzy.

Hall, N., & Taplin, R. (2008). Room for climate advocates in a coal-focused economy? NGo influence on Australian climate policy. *Australian Journal of Social Issues 43*, 359–379.

Hess, D. J. (2018). Energy democracy and social movements: A multi-coalition perspective on the politics of sustainability transitions. *Energy Research & Social Science, 40*, 177–189.

Klein, N. (2014). *This Changes Everything: Capitalism vs. the Climate*. New York: Simon & Schuster.

Mark, J. (2013). Conversation: Naomi Klein. *Earth Island Journal* (Autumn). http://bit.ly/18R9oZ2.

McKibben, B. (2013). Beyond baby steps: Analysing the cap-and-trade flop. *Grist*. http://bit.ly/1dGK3sU.

Murphy, G. (2005). Coalitions and the development of the global environmental movement: A double-edged sword. *Mobilization, 10*, 235–250.

Nunes, R. (2009). What were you wrong about ten years ago? *Turbulence, 5*, 38–39.

Parag, Y., & Janda, K. B. (2014). More than filler: Middle actors and socio-technical change in the energy system from the middle-out. *Energy Research & Social Science, 3*, 102–112.

Plumer, B. (2013, January 16). *Why has climate legislation failed?* An interview with Theda Skocpol. *The Washington Post*. http://wapo.st/1jFMBUA.

Polleta, F., & Jasper, J. M. (2001). Collective identity and social movements. *Annual Review of Sociology, 27*, 283–305.

Rosewarne, S., Goodman, J., & Pearse, R. (2014). *Climate Action Upsurge: The Ethnography of Climate Movement Politics*. Abingdon, Oxfordshire, UK, and New York: Routledge.

Routledge, P. (2012). Translocal climate justice solidarities. In J. S. Dryzek, R. B. Norgaard, & D. Schlosberg (Eds.), *The Oxford Handbook of Climate Change and Society*. Oxford, UK: Oxford University Press.

Saunders, C. (2008). Double-edged swords? Collective identity and solidarity in the environmental movement. *The British Journal of Sociology, 59*, 227–253.

Saunders, C. (2009). It's not just structural: Social movements are not homogeneous responses to structural features, but networks shaped by organisational strategies and status. *Sociological Research Online, 14*, 1–16.

Saunders, C. (2013). Insiders, thresholders, and outsiders in west European global justice networks: Network position and modes of coordination. *European Political Science Review, 2*, 167–189.

Schock, K. (2005). *Unarmed Insurrections: People Power Movement in Nondemocracies*. Minneapolis: University of Minnesota Press.

Seyfang, G., Park, J. J., & Smith, A. (2013). A thousand flowers blooming? An examination of community energy in the UK. *Energy Policy, 61*, 977–989.

Skocpol, T. (2013). *Naming the Problem*. Cambridge, MA: Harvard University Symposium on the politics of America's fight against global warming. http://bit.ly/Pc8W3O.

Snow, D. A., Rochford, B., Worden, S. K., & Benford, R. D. (1986). Frame alignment processes, micromobilization, and movement participation. *American Sociological Review, 51*, 464–481.

Tarrow, S. (1998). *Power in Movement: Social Movements and Contentious Politics*. New York: Cambridge University Press.

Tarrow, S. (2005). *The new transnational activism*. Cambridge, UK: Cambridge University Press.

Tilly, C. (1995). *Popular Contention in Great Britain, 1754–1834*. Cambridge, MA: Harvard University Press.

Tilly, C. (2002). *Stories, Identities and Political Change*. New York: Rowman and Littlefield.

Tilly, C. (2008). *Contentious Performances*. Cambridge, UK: Cambridge University Press.

United States Climate Action Partnership (USCAP). (2009a). About us, http://bit.ly/1yFcgbu.

USCAP. (2009b). About our members, http://bit.ly/1BTJ8ja.

Van der Schoor, T., Van Lente, H., Scholtens, B., & Peine, A. (2016). Challenging obduracy: How local communities transform the energy system. *Energy Research & Social Science, 13*, 94–105.

CHAPTER 6

Interacting

Abstract Interacting is a shorthand for activists reaching out to a large proportion of the population. Numbers remain imperative for effective mobilisation; thus, interactions between people of like minds and other commonalities remain an imperative for climate actions. With low media attention and unfriendly media stance over climate issues, this chapter asks how do contemporary action groups achieve effective public communication, how can these strategies be translated into climate actions, what are the strengths and limitations of new media, and how do contemporary action groups use them effectively. Furthermore, the chapter asks how can face-to-face interactions be balanced with the strategic use of social media alongside deliberative exercises.

Keywords Climate communication · Media · Social media
Deliberation

Framing an issue to make it emotive, memorable, with a clear wrong, and a convincing solution that requires the individual to take action. It needs to be simple enough to be easy to communicate quickly to busy people with no prior interest, but the quick and simple version needs to avoid distortions.

© The Author(s) 2019   73
L. L. Delina, *Climate Actions*,
https://doi.org/10.1007/978-3-319-91884-6_6

—Respondent from a UK, faith-based, international development group formed in 1968 working on long-term development and disaster relief.

A successful movement is often hinged on its capacity to achieve widespread publicity (Moyer et al. 2001; Delina et al. 2014); hence the need for stronger media engagement. An important example of the strong correlation between successful mobilisation and extensive media reporting of past social mobilisations was the coverage of Gandhi's march to Dandi where he made salt for himself and the violent dispersal of peaceful protesters at Dharasana saltworks, weeks after said march. These two events received international media attention. The latter, however, became the defining moment for the Indian independence movement because of the worldwide publication of journalist Webb Miller's account of the event.[1] The extensive media coverage of the violence at Dharasana paved the way for the rapid mobilisation of both local and international support for Indian independence (Weber 1997).

History is rife with many other moments showing how media helped in tipping the scale in favour of social movements. Most respondents strongly agree that extensive media coverage ($N=42$, median$=4$, Interquartile range (IQR)$=1$[2]) and critical social problem expositions in popular media ($N=39$, median$=4$, IQR$=0$) make mobilisations easier and quicker. Twenty-five of them or 60% state their campaigns have become 'effective' with extensive media coverage, while fourteen or 33% affirm that this attention had led to 'very effective' campaigns. When mainstream media provides extensive coverage, most respondents feel they have been successful ($N=42$, median$=4$, IQR$=1$). Not all mobilisations, however, can be granted identical or appropriate media exposure and coverage. The climate action movement has been presented with this dilemma.

Empirical examples have shown how low media attention and unfriendly media stance over climate issues have been prominent in high emission countries such as Australia and the USA (Bacon 2013; Bacon and Nash 2012; Feldman et al. 2011). In these places and elsewhere, climate change reporting continues to be narrowly framed either as an impending or an emerging environmental problem with serious

---

[1] Miller's account was even read in the US Congress and admitted to its official records.

[2] See Appendix B for notes on statistical treatment and reporting.

consequences in the future instead of as an ongoing and present social, economic, and political issue (Boykoff and Boykoff 2004). Positive stories about climate actions, such as sustainable energy transitions, also often do not merit as much airtime compared to stories on violence, wars, and political bickering.

In cases when climate change is reported, many mainstream media outlets, particularly in the USA and Australia, have also shown bias against climate actions (Feldman et al. 2011; Bacon and Nash 2012). When interviewing climate scientists, for example, equal time is often afforded to non-scientist deniers of climate science. Rupert Murdoch's *Fox News*, the *Wall Street Journal*, and the *New York Post*, for example, have provided opportunities for representatives of conservative think tanks, climate science deniers, and contrarian scientists to consistently ridicule and assault climate change, the International Panel on Climate Change, and climate scientists (Dunlap and McCright 2011). In its report on the coverage of climate change in Australian print publications, the Australian Centre for Independent Journalism has demonstrated the persistent misleading and confusing news reports about scientific findings on climate science, especially in Murdoch-owned presses (Bacon 2013). Misinformation and doubt-sowing reporting in Australia includes *The Australian*, the country's only national newspaper for a general audience (Bacon 2013). Not surprisingly, Murdoch's *The Daily Telegraph* has even subjected this author's work to a derogatory attack (see Thomas 2013).

The fossil fuel regime, which has extended its reach and influence in mainstream media has, indeed, been impacting the framing of climate science and climate actions, as demonstrated for instance by the powerful Australian coal industry (Bacon and Nash 2012). Rupert Murdoch's *News Corporation* through its tabloid *The Daily Telegraph* in Sydney and *The Courier Mail* in Brisbane carry news stories based on the primary assumption that the growth and expansion of coal industry is favourable for Australia (Bacon and Nash 2012) and, therefore, has to be protected. The attitude of Australian media towards climate change remains, at best, tilted towards protecting the country's coal industry.

The absence of media coverage of the climate action movement by mainstream broadcasters was also evident during the People's Climate March. Despite the event's huge turnout, popular US media outlets failed to cover the campaign (Johnson 2014). While *Al Jazeera America* provided airtime reporting on the March (Mirkinson 2014), only *DemocracyNow!*, an independent media organisation, broadcasted the entire event live for three hours (*DemocracyNow* 2014). *NBC Nightly News* was the sole primetime news programme to air a segment about the march; ABC used approximately 23 seconds on the topic in its own newscast; while, CBS devoted zero seconds on it (Mirkinson 2014).

Communicating climate actions with the hope of mobilising a movement via traditional media outlets, thus, may not be the most effective option for climate action campaigners. If individuals and communities do not comprehend either the threats of climate change or its solutions or both, they cannot be expected to endorse changes to mitigate it. Since unsupportive media, according to 33 respondents or 70%, pose a barrier to campaigning, climate actions must be designed in ways that they can reach various audiences without much reliance on traditional media.

'Non-traditional' means, modes, and channels of communication provided by online media and social media may boost audience reach of climate actions. In the digital era, Internet-based news sources have charted their own virtual territories and claimed audiences previously held by popular media institutions. Indeed, popular media agencies are already grappling with lost TV viewership, radio listenership, and magazine and newspaper readership (Orr 2013; Slaughter 2005). SAO07[3] recognises this opportunity saying: "Having strong direct [communication] to supporters is essential, you can't rely on the media alone. Magazines, postcards, films, email, social media and a website are all very useful."

A strategy that takes advantage of tapping multiple media also makes sense (Agyeman et al. 2007; Featherstone et al. 2009; Leiserowitz et al. 2009) given the varied preferences of audiences for climate actions. A study from Yale and George Mason Universities about climate communications in the United States found that people have indeed varied selections as to their preferred news sources (Leiserowitz et al. 2009). While majority of the study's respondents or about 59% prefer the television as their most preferred source, those who prefer the Internet had risen to 22% in 2008 (Leiserowitz et al. 2009). This figure must have risen significantly in recent years, especially with the role of social media gaining prominence in American society. Communicating climate actions, therefore, will now ideally involve pairing audiences with specific means,

---

[3] See Table C.1 for the description of the social action groups.

modes, or channels of communication beyond those provided by traditional media outlets (Bostrom et al. 2013; Moser 2007a, b).

The Internet provides inexpensive media to distribute climate action messages to anyone with access to it. Alongside this, the emergence of Internet-assisted national and international climate action networks have also been allowing for more decentralised, self-replicating, and even self-correcting alternative platforms (Orr 2013). These could be beneficial for communicating climate actions since they are faster and cheaper than those provided by traditional systems (Slaughter 2005). A majority of our survey respondents strongly agree on the use of Internet-based communication tools in campaigns ($N=46$, median$=4$, IQR$=0$). SAE04, for example, has been "maintaining a website with resources such as handouts, educational materials, bumper stickers, pins and lawn signals, for groups and individuals to use as a key campaign strategy."

Using new media is not new in large-scale mobilisations. Beginning in the turn of the new millennium, the power of communications technology have already mobilised many social action campaigns. The Philippines, for instance, became the location of the first mobile phone-propelled social mobilisation in history. During the 2001 EDSA People Power Revolution, the second large-scale nonviolent movement that overthrew a sitting President in that country, Filipinos mobilised themselves through cellphone text messages (Shirky 2011). This strategy has been adopted many times since, including in the 2004 ousting of Spanish Prime Minister Hose Maria Aznar (Suarez 2006) and the 2009 toppling of Moldova's ruling Communist Party (Mungiu-Pippidi and Munteanu 2009). Some contemporary social action groups also continue to rely on SMS-based mobilisation; however, the opinions of our survey respondents are divided as to its effectiveness. Ten respondents say this mode is 'effective' ($N=9$, 50%) and 'very effective' ($N=1$, 6%), but a roughly close number of respondents see this approach as 'ineffective' ($N=7$, 39%) and 'very ineffective' ($N=1$, 6%).

Smart phone innovations, particularly the emergence of 'apps' functionality in the early 2010s, have effectively changed the information and communications landscape. Apps have largely been attributed as enablers of mobilisations of the 'Arab Spring,' including the large-scale assemblies at Tahrir Square in Cairo (Stepanova 2011). Many social action groups use Internet-based media platforms, such as Twitter, Facebook, Instagram, Linked-In, YouTube, Vimeo, and Vine, among others, in their mobilisation work. Thirty-four respondents or 74%

suggest that social media mobilisation has been 'effective,' while nine or 20% recognise it as a 'very effective' approach.

The utility of social media as a mobilisation tool was also evident during the 2014 People's Climate March. With approximately 630,000 social media posts generated during the event, its organisers claimed that social media was the March's important communication channel (People's Climate March 2014). The utility of social media also is now being extended to document and coordinate real-world action, a province previously and almost exclusively held by media entities. But social media mobilisations also have records of failure. In 2009, for example, the Green Movement uprising in Iran, which used every available social media tool, especially Twitter, to hold a post-election protest, ended with a violent crackdown (Karagiannopoulos 2012; Kurzman 2012). In 2010, the Red Shirt uprising in Thailand, which also was mobilised largely through social media, similarly ended in violent dispersals (Shirky 2011).

The digital revolution has brought with it profound and complex issues to contemporary societies. Most relevant to mobilisations is the degree to which authenticity in movement participation can be secured by and with the new technology. By authenticity, I mean the degree to which publics actually engage themselves in actual, reasoned discourse rather than the mere strategic pursuit of interest or symbolic participation. For example, electronic actions, such as through pushing a button to add one's name to an e-petition or sending an email to a Member of Parliament, are largely unthinking, isolated, and one-way (Schlosberg and Dryzek 2002). These approaches easily deplete social capital that used to animate a vibrant social action movement. The rise of untrustworthy information, or fake news, in the digital era also exposes the vulnerabilities of many contemporary societies. The preponderance of fake news, largely and almost exclusively purveyed in social media, has become contemporary societies' major challenge. As a metaphor, fake news has allowed sewage to flow alongside potable water through the same distribution pipe. There is no panacea to this problem but reflexivity remains a key approach. With it, campaigners must underline the fact that we do not simply take modern information technologies as given but we must also question how technology itself helps construct or reconstruct society and its orderings.

Social media has already shown its potential in organising and mobilising individuals and groups into climate actions, but it still takes face-to-face interactions to construct authentic, unfiltered, and engaged climate actions. This is essentially true in the age of fake news. Our survey respondents strongly agree that face-to-face conversations and meetings remain 'effective' communication channels ($N=16$, 34%), with thirty-one of them (66%) saying it is 'very effective' (cf. Chapter 1). What is clear in these face-to-face interactions, however, is to make "a very clear call to action...where after each meeting the attendees are asked to organise their own" (SAE04, SAE05, SAE06, and SAO07). These conversations "must also be clearly facilitated using some creativity to tell the story of people affected by climate change" (SAE03), "geared to where the audience is at" (SAE37; cf. Chapter 2), and to communicate the climate action agenda, including "stories of communities that have responded creatively to the realities of climate change" (SAE03; cf. Chapter 4). As discussed in previous chapters, there are plenty of ways to achieve this. SAE13, in addition, suggests "using analogies with similar issues that do have critical problems such as comparing the fossil fuel industry with tobacco or asbestos industry." SAE24 notes: "reframing questions to take into account 'green fatigue' such as 'do you support renewable energy?' instead of 'do you believe in climate change?'

Even in the age of social media and fake news, many respondents ($N=43$) also strongly agree that traditional, 'offline' materials such as newsletters and other printed materials still have roles to play in communicating about campaigns (median $=4$; IQR$=0$). The use of newsletters and other printed materials have indeed remained high in their communication strategies with 33 respondents or 77% stating it is an 'effective' approach. In addition to these channels, respondents also suggest investing in other channels such as placing ads in public places and public transport, and in "using the arts, for example, theatre, music (SAE30), literature, film and visual art (SAE03, SAE18, SAE21, SAO03), stunts and performances (SAE12), and other [related] events" (SAE36).

While climate actions have not been in the crosshairs of many traditional media outlets, twenty-four respondents or 56% still look at them as relevant partners. Strategies to engage the media include "sending open letters signed by senior religious leaders; writing letters to editors" (SAE31); and "writing occasional articles, op-eds, and media releases; and placing ads in newspapers." SAE33 also suggests "producing media stories featuring well-liked and well-respected community leaders who will publicly state their concerns about climate change and the need for stronger action." SAE25 narrates how they found the model for organising developed by Marshall Ganz to be helpful:

> Using a Ganz model for building a campaign that has gradual steps of increasing engagement, having the grass roots build [a/the] movement so that we can pull in more people at every step and build bigger events and actions. Then the media begins to take notice and cover us, and we often get the prominent person endorsement that will boost our campaign further. Sometimes the action puts the spotlight on the issue and the rest follows. We have had good success with this, especially in our oil-by-rail resistance campaign, which began ten days prior to the Lac Megantic rail disaster.[4]

Broadly, communication strategies reduce constraints on participation, which include public perceptions of the risk of getting involved in social actions (Chenoweth and Stephan 2011). Critical mass theories of social action, which state that audiences base their perceptions of opportunities to get engaged on patterns of the activities by social action campaigners (Kurzman 1996), are also relevant here. There is no hard and fast rule on how perceptions are changed but some public engagement exercises provide options to achieve this level of engagement.

Given the heterogeneous character of climate actions and their participants, deliberative exercises may be employed to mobilise people into climate actions. Deliberative exercises are processes of iterative knowing, probing, and solving that consider the variety of opinions and suggestions from a heterogeneous mix of participants (Chomsky 2012; Carson 2011, 2010; Dryzek 2009, 2000; Fung 2008; Gutmann and Thompson 2004). Deliberations, not debates, characterise these public exercises. Their outputs are not necessarily a consensus but diversified sets of policy or action recommendations that reveal the plurality of intentions of its participants while respecting the multiple types of knowledge and expertise they possess and bring to the exercises (Carson 2011; Dryzek 2000). These facilitated deliberative exercises encourage sharing, testing, and weighing of various options for climate actions, thus allowing people to develop a deeper understanding, not only of the climate issues, but also their possible solutions (Carson 2011; Dryzek 2009; Stirling 2008; Nagda 2006; Jasanoff 2004). Some real-world examples of these exercises have persisted and registered successes. These include, for example, deliberations among the indigenous Sami populations in Finnmark

---

[4]On 6 July 2013, a freight train carrying crude oil derailed in the town of Lac Megantic in Quebec, Canada resulting in a fire and an explosion that killed 42 people.

who deliberated on the location and speed of a wind farm construction (McCauley et al. 2015). In achieving authentic deliberative exercises for climate actions, professionally facilitated meetings are necessary where people can deliberate the climate issue and identify a variety of possible solutions and approaches that are relevant to them (cf. Chapter 1). These solutions can range from community sustainable energy to local energy efficiency (cf. Chapter 4), among many other possible pathways of climate actions. To magnify impact, these pockets of deliberations for climate actions must be ideally linked in a larger deliberative system (Niemeyer and Jennstål 2018; Mansbridge et al. 2012; cf. Chapter 5).

In closing, the availability of diverse, readily available modern communication means, modes, technologies, and channels places climate action campaigners in an advantaged position. They are now provided with these tools to carry out not only campaigning but also documentation and coordination of the many possible climate actions. Despite this, social media—the most favoured modern communication technology available to campaigners—are two-edged swords. On one hand, they provide new opportunities for securing greater public engagement; on the other hand, they lend less authenticity to the required climate actions and are prone to fake news. Balancing the use of social media with face-to-face engagement, thus, remains vital in climate mobilisation. To that end, more deliberative exercises that secure authentic, influential, and inclusive public engagement on climate actions offers a way forward.

References

Agyeman, J., Doppelt, B., Lynn, K., & Hatic, H. (2007). The climate-justice link: Communicating risk with low-income and minority audiences. In S. C. Moser & L. Dilling (Eds.), *Creating a Climate for Change: Communicating Climate Change and Facilitating Social Change* (pp. 119–138). Cambridge, UK: Cambridge University Press.

Bacon, W. (2013). *Climate Science in Australian Newspapers.* Sydney, Australia: The Australian Centre for Independent Journalism.

Bacon, W., & Nash, C. (2012). Playing the media game: The relative (in)visibility of coal industry interests in media reporting of coal as a climate change issue in Australia. *Journalism Studies, 13,* 243–258.

Bostrom, A., Böhm, G., & O'Connor, R. E. (2013). Targeting and tailoring climate change communications. *WIREs Climate Change, 4,* 447–455.

Boykoff, M. T., & Boykoff, J. M. (2004). Balance as bias: Global warming and the US prestige press. *Global Environmental Change, 14,* 125–136.

Carson, L. (2010, September 9). Growing up politically: Conducting a national conversation on climate change. *Australian Policy Online.* http://bit.ly/ONNRaH.

Carson, L. (2011). Dilemmas, disasters and deliberative democracy: Getting the public back into policy. *Griffith Review Edition 32: Wicked Problems, Exquisite Dilemmas, 32*(Winter), 25–32.

Chenoweth, E., & Stephan, M. J. (2011). *Why Civil Resistance Works: The Strategic Logic of Nonviolent Conflict.* New York: Columbia University Press.

Chomsky, N. (2012). *Occupy.* New York: Zuccotti Park Press.

Delina, L., Diesendorf, M., & Merson, J. (2014). Strengthening the climate action movement: Strategies from histories. *Carbon Management 5,* 397–409.

Democracy Now! (2014, September 21). *Special 3-Hour Broadcast of the People's Climate March.* http://bit.ly/1tHsEFF.

Dryzek, J. (2000). *Deliberative democracy and beyond: Liberals, critics, contestations.* Oxford, UK: Oxford University Press.

Dryzek, J. (2009). Democratization as deliberative capacity building. *Comparative Political Studies, 42,* 1379–1402.

Dunlap, R. E., & McCright, A. M. (2011). Organized climate change denial. In J. S. Dryzek, R. B. Norgaard, & D. Schlosberg (Eds.), *The Oxford Handbook of Climate Change and Society* (pp. 144–160). Oxford, UK: Oxford University Press.

Featherstone, H., Weitkamp, E., Ling, K., & Burnett, F. (2009). Defining issue-based publics for public engagement: Climate change as a case study. *Public Understanding of Science, 18,* 94–101.

Feldman, L., Maibach, E. W., Roser-Renouf, C., & Leiserowitz, A. (2011). Climate on cable: The nature and impact of global warming coverage on Fox News, CNN, and MSNBC. *The International Journal of Press/Politics 17,* 3–32.

Fung, A. (2008). Democratizing the policy process. In R. Goodin, M. Rein, & M. Moran (Eds.), *The Oxford Handbook of Public Policy* (pp. 669–688). Oxford, UK: Oxford University Press.

Gutmann, A., & Thompson, D. (2004). *Why Deliberative Democracy?* Princeton, NJ: Princeton University Press.

Jasanoff, S. (2004). Science and citizenship: A new synergy. *Science and Public Policy, 31,* 90–94.

Johnson T. (2014). *Sunday News Shows Ignore Historic Climate March.* Media Matters for America. http://bit.ly/1wWLr27.

Karagiannopoulos, V. (2012). The role of the internet in political struggles: Some conclusions from Iran and Egypt. *New Political Science, 34,* 151–171.

Kurzman, C. (1996). Structural opportunity and perceived opportunity in social movement theory: The Iranian Revolution of 1979. *American Sociological Review 61,* 153–170.

Kurzman, C. (2012). The Arab Spring: Ideals of the Iranian Green Movement, methods of the Iranian Revolution. *International Journal of Middle East Studies 44,* 162–165.

Leiserowitz, A., Maibach, E., & Roser-Renouf, C. (2009). *Global Warming's 'Six Americas' 2009: An Audience Segmentation Analysis.* New Haven, CT: Yale Project on Climate Change: School of Forestry and Environmental Studies, Yale University; Fairfax, VA: Center for Climate Change Communication and George Mason University.

Mansbridge, J., Bohman, J., Chambers, S., Christiano, T., Fung, A., et al. (2012). A systemic approach to deliberative democracy. In J. Parkinson & J. Mansbridge (Eds.), *Deliberative Systems: Deliberative Democracy at the Large Scale* (pp. 1–26). Cambridge, UK: Cambridge University Press.

McCauley, D., Rehner, R. W. M., & Pavlenko, M. (2015). Assessing the justice implications of energy infrastructure development in the Arctic. In R. J. Heffron & G. Little (Eds.), *Delivering Energy Law and Policy in the EU and US.* Edinburgh, UK: Edinburgh University Press.

Mirkinson, J. (2014, September 22). TV news misses yet another opportunity to cover climate change. *The Huffington Post.* http://huff.to/1sUOqkX.

Moser, S. C. (2007a). Communication strategies to mobilize the climate movement. In J. Isham & S. Waage (Eds.), *Ignition: What You Can Do to Fight Global Warming and Spark a Movement* (pp. 73–95). Washington, DC: Island Press.

Moser, S. C. (2007b). In the long shadows of inaction: The quiet building of a climate protection movement in the United States. *Global Environmental Politics, 7,* 124–144.

Moyer, B., McAllister, J., Finley, M. L., & Soifer, S. (2001). *Doing Democracy: The MAP Model for Organizing Social Movements.* Gabriola Island, BC: New Society Publishers.

Mungiu-Pippidi, A., & Munteanu, I. (2009). Moldova's "Twitter revolution". *Journal of Democracy, 20,* 136–142.

Nagda, B. R. A. (2006). Breaking barriers, crossing borders, building bridges: Communication processes in intergroup dialogues. *Journal of Social Issues, 62,* 553–576.

Niemeyer, S., & Jennstål, J. (2018). *Scaling up deliberative effects—Applying lessons of minipublics.* Oxford, UK: Oxford University Press Handbook of Deliberative Democracy, Oxford University Press.

Orr, D. W. (2013). Governance in the long emergency. In E. Assadourian & T. Prugh (Eds.), *State of the World 2013: Is Sustainability Still Possible?* (pp. 279–291). Washington, DC; Covelo; and London, UK: Island Press.

People's Climate March. (2014). *Wrap Up.* http://bit.ly/1qxNEZt.

Schlosberg, D., & Dryzek, J. S. (2002). Digital democracy: Authentic or virtual? *Organization & Environment, 15,* 332–335.

Shirky, C. (2011). The political power of social media: Technology, the public sphere and political change. *Foreign Affairs, 90,* 28–41.

Slaughter, A.-M. (2005). *A New World Order.* Princeton, NJ: Princeton University Press.

Stepanova, E. (2011). *The role of information communication technologies in the "Arab Spring": Implications beyond the region.* Washington, DC: *PONARS Eurasia*, the George Washington University, and Elliott School of International Affairs. http://bit.ly/1L9ceOg.

Stirling, A. (2008). 'Opening up' and 'closing down': Power, participation and pluralism in the social appraisal of technology. *Science, Technology, and Human Values, 33,* 262–294.

Suarez, S. L. (2006). Mobile democracy: Text messages, voter turnout and the 2004 Spanish general election. *Representation, 42,* 117–128.

Thomas, T. (2013, December 19). A room full of eco-idiots. *The Daily Telegraph.* http://bit.ly/1uJcBFk.

Weber, T. (1997). *On the Salt March.* New Delhi, India: HarperCollins.

CHAPTER 7

Conclusion

Abstract Producing an alternative ordering of contemporary human societies—one that is more sustainable, fair, and just—needs to arise from a rupture of fossil-based order. This requires multiplied climate actions of a strengthened climate action movement. Doing this is not a utopian work but requires series of and parallel efforts to relate actions with people's everyday lives and aspirations, to deliver spatially and temporally proximate climate action messages, to develop a clear vision of the alternatives, to web these heterogeneous actions, and to interact with publics using multiple tools. And while success could not be guaranteed, histories have shown that a focused approach on these transformative mechanisms could mobilize peoples.

Keywords Paris Agreement · Climate actions · Climate activism Mobilisation

The year 2015 can be considered as a watershed year for global climate action. In December of that year, governments agreed in Paris to curtail future emissions. But the Agreement, according to climate scientist Kevin Anderson, is a 'fantasy' in that this 32-page document has "no meaningful substance in delivering on the temperature obligations... says nothing about de-carbonization, and doesn't even mention fossil fuels!" (Anderson and Nevins 2016: 210). What this tells us is that the work for climate actions does not end in Paris; rather the Paris

L. L. Delina, *Climate Actions*,
https://doi.org/10.1007/978-3-319-91884-6_7

85

Agreement provides a propitious starting point for *real* climate actions to occur—and for strengthening them. With the climate change challenge remaining a standing goal for humanity, climate actions must adequately address the urgency of the transition to a sustainable economy, sustainable consumption and production, and climate equity, democracy, and justice. This book enriches the discussion on climate actions, in particular the development of mechanisms for making them effective.

The broad, dynamic, multi-scale, and heterogeneous climate action movement has truly achieved some pockets of successes in that it was able to produce outward-oriented forms of direct actions such as protests and demonstrations, most notably the People's Climate March and fossil fuel divestment. Yet its work has received little political traction (vis-à-vis the required transformation). Fossil fuels continue to be extracted and burned. Community energy that activists have been strongly advocating is taking time to materialise in the large scale. The failures to replicate innovations we first saw in the German *Energiewende* have become commonplace. Despite this frustration, locally organised and mobilised climate actions must be sustained. They remain our best shot to challenge normal, routine politics. The movement needs now to carefully rebalance its strategies, tactics, and campaigns. Paramount to this is to embed more prefigurative type of public engagement. The climate action movement can accomplish this goal in many ways, including advancing more and diversified actions such that they are not seen solely and strictly as exercises geared for climate ends, but also contingent to the personal aspirations, priorities, hopes, and dreams of whoever their intended audiences are. We are in dire need for deeper transformations towards not only a climate safe society but a steady state and post-growth economy. Reaching these goals require that climate actions are heterogeneous and are traversed in multiple pathways—all at the same time.

What is key in climate actions is to focus on peoples—and to solicit a much stronger engagement from them. Eliciting stronger public engagement towards meaningful, clear, and beneficial climate actions remains an important modus operandi to achieve transformations. However, these processes require a reflexive menu of transformative strategies that respond to, accept, and respect the extreme heterogeneity in experiences, knowledge, wealth, and wants of peoples whom we must mobilise across our communities, politics, time, and space. Truly, these mobilisations must also be built and sustained from the ground-up. The contemporary experiences of many social action groups, this book illustrates, proved to be dynamic wellsprings by which we can develop and strengthen the transformative mechanisms needed to strengthen both outward-focused and prefigurative types of climate actions.

These mechanisms are not a simple "to-do" list; they are based on a theoretical infrastructure provided by the social movement literature and supported by real-world experiences of those who have been in the barricades. Dissent, as has been argued here, aims at breaking norms and is performed in multiple ways. In this multiplicity, this book has argued for alternative orderings that (1) highlight face-to-face interactions that would turn everyday talk into effective climate actions (*Relating*); (2) recognise that reflexive framing and psychological stimuli can affect people's attitudes towards climate actions (*Messaging*); (3) work towards a new collective identity through a clear vision of alternative sustainable futures for all (*Visioning*); (4) establish strong linkages across heterogeneous climate actions (*Webbing*); and (5) engage with multiple frames and media for communicating climate actions (*Interacting*). The overall thrust of these mechanisms is to produce an alternative ordering of contemporary human societies—one that arises from a rupture of the fossil-based order in favour of more sustainable, fair, and just systems. To be clear, however, this is not a utopian work.

Real-world examples of social mobilisations suggest these paths are winding and the processes to traverse them are contentious. The climate action movement—involving individuals, households, small groups, and communities, acting either as pockets of innovation, yet ideally linked in an optimistic, durable, and doable vision of the future—is presented with a vast opportunity to shatter the familiar landmarks of contemporary climate actions. This requires that the usual order is revealed through clear communication of who the opponents are (including how they imagine the future to be based on their paradigms and actions) and what the alternative future could be (including how we can reach that future in a just, sustainable, and fair manner). Unsettling the fossil fuel-based regime and destabilising their foundations, indeed, does not only mean making these existing orders legible to all peoples but it is also about embedding alternative narratives and imaginations of what a durable and sustainable future *ought* to look like. The climate action movement, thus, in its work and climate actions, must demonstrate that social orderings can be made differently.

The strategies presented in this book are by no means innovative. I confess they are not even new. Also, the effectiveness of climate actions could never be guaranteed by employing these strategies. Making more climate actions seem to be ideal but the fact of the matter is there are more failures in local mobilisations than there are successes. What I want to convey, therefore, is that the mechanisms for social transformations I have described in this book have remained consistent with the theoretical and conceptual frameworks established in the scholarly literature on social movements and the empirically understood structure of effective modern social action campaigns. I hope this can provide the climate action movement a reflexive tool by which it can review its contemporary work.

The climate action movement needs not reinvent the wheel. Instead, it must continue and sustain the creative processes for critical reflection towards the sets of tools it may already have and the experimentation with new ways to critically act in the world. These moments of reflection within the movement are critical as it allows a collective to build its strategies in the struggle to mobilise publics around the necessity of urgent climate actions. To that end, I see the raison d'être for more and strengthened actions spurred in the local level and the need to link these pockets in a pluriverse that actually makes the climate action movement. I hope this book has provided an opportunity for examining these strategies especially in this critical juncture in our history. Collective climate actions are the antipode to those in power who continue to elect on extremely tentative steps in dealing with the threat to the future habitability of our common home. These *Climate Actions* must be strengthened and multiplied.

REFERENCE

Anderson, K., & Nevins, J. (2016). Planting seeds so something bigger might emerge: The Paris Agreement and the fight against climate change. *Socialism and Democracy 30*, 209–218.

Appendix A: Survey Questionnaire

Part 1: About your group

- Name of your group, including acronym or abbreviation.
- Location.
- Year your group was created/founded.
- Membership size.
- Principal category/group orientation: environmental; social justice and/or human rights; other, please specify.
- Specific social action agenda (e.g. anti-nuclear; religious/faith; civil rights; climate action; Indigenous Peoples; labour).
- What does your group specifically do?
- Coverage/extent of work: local/community based; state-wide/province-wide; national; international; online.
- Is your group a member of a network, alliance, or coalition? If yes, please specify.

Part 2: Effective campaign strategies

- Rate the effectiveness of the following strategies for accomplishing social change.
  Scale: No experience; Very ineffective; Ineffective; Effective; Very effective.

- Education, information, awareness campaigns
- Non-violent direct action
- Both.
- Following on question above, what other approaches can you suggest in making social action campaigns effective? Please elaborate on your response.
- Rate the effectiveness of the following strategies for getting public support and engagement.
  Scale: Scale: No experience; Very ineffective; Ineffective; Effective; Very effective.
  - Connecting campaigns with moral values.
  - Involving or engaging prominent people.
  - Extensive media coverage.
  - Presence of shocking incident or event that dramatically highlights a critical social problem.
  - Exposing the critical social problem in popular media.
  - Careful planning.
  - Affiliating in an alliance, network, or coalition.
  - Targeting campaigns towards existing social groups (e.g. faith groups, cultural groups, professional groups).
  - Engaging a public who share similar interests, faith, profession, age brackets, or community or area.
  - Engaging with friends, family, or relatives.
  - Engaging in joint activities even though they dare none of the above characteristics.
- Following on question above, what other strategies can you suggest in soliciting public support and engagement? Please elaborate on your response.
- Rate the importance of the following moral values in your campaigns.
  Scale: No experience; Not important; Somewhat important; Important; Very important.
  - Nonviolence.
  - Social justice.
  - Right to security.
  - Intergenerational equity.
  - Conservation of biodiversity and ecological integrity.

- Basic human rights.
- Ecocentric rather than anthropocentric position.
- What other moral values can be invoked? Please elaborate on your response.
- In the absence of a critical social problem highlighted in the media, or exposed by a shocking incident, or supported by prominent people, what other strategies can you suggest?
- Outcomes from affiliating with a network or an alliance.
  Scale: Strongly disagree; Disagree; Agree; Strongly Agree.
  - Our group can create greater public impact since we are now part of a bigger group.
  - We feel we are now a stronger force.
  - We can combine our group's resources with that of other groups.
  - We feel we have lost part of our independence.
- What other outcomes have you experienced from being in a network, coalition or alliance? Please describe these sentiments.
- Challenges and barriers to effective campaigns?
  - Lack of funds and other resources.
  - Unsupportive media.
  - Absence of prominent persons in campaigns.
- What other challenges have you experienced? Please elaborate on your response.

Part 3: Ensuring public engagement

- Rate the effectiveness of the following modes of communication in ensuring public engagement as defined above.
  Scale: No experience, Very ineffective; Ineffective; Effective; Very effective.
  - Face-to-face conversations.
  - Social media.
  - SMS or text messages.
  - All or combination of the above.
- What other communication strategies to ensure public engagement can you suggest? Please elaborate on your response.

Appendix B: Notes on the Research Instrument, Statistical Treatment, and Reporting

The University of New South Wales' Built Environment Human Research Ethics Advisory Panel approved the research instrument on its meeting of 28 August 2014. The survey was administered between 28 August and 16 September 2014. The Boston University Charles River Campus Institutional Review Board, through its Protocol No. 4691X, approved the same survey instrument on 12 December 2017. It was administered between 13 December 2017 and 13 January 2018.

To assist the respondents with their choices, they were provided a range of choices through a four-level Likert scale, plus an option for 'no experience.' The scale does not have a 'neutral' option to reduce social desirability bias, or the tendency of the respondent to answer questions in a manner that will be viewed favourably by others. Study shows that social desirability bias is prone among scales with neutral options and arises from respondents' wanting 'to please the interviewer or appear helpful or not be seen to give what they perceive to be a socially unacceptable answer' (Garland 1991: 66–69).

To allow for free-flow of additional insights, the instrument also contains opportunities for respondents to voice an opinion, to expand on their answers, or even rebut the predetermined choices. The instrument was designed in this way to allow for some level of interaction. Another point to make about the scale is with regards to not using a Likert scale that contains large number of items, for example, seven,

L. L. Delina, *Climate Actions*,
https://doi.org/10.1007/978-3-319-91884-6

nine, or ten choices. Although a large scale might capture a variety of responses and, therefore, might be more sensitive to nuanced positions, having a large number of choices can become too cumbersome to use and any additional benefits are cancelled by respondent fatigue (Ben-Nun 2008). Therefore, to safeguard reliability of response from plummeting, a shorter four-scale Likert scale, without a neutral option, was used.

To further minimise bias, no question was raised involving climate change or climate actions; rather, the topics revolved around broader issues that have universal application to social action regardless of campaign topic. By omitting climate change-related questions, the survey instrument allowed respondents, especially those who are neither environment nor climate-oriented, to express themselves freely without making any climate-related assumptions.

The statistics presented in this book generally represents verbal, not numerical, statements. To appreciate them as qualitative statements, similar items were clustered and the responses to these questions were compared and merged. Qualitative responses gathered from responses to follow-up questions have, thus, provided supplements that strengthened the statistical responses. Only when responses to these follow-up questions become broadly consistent with the medians and Interquartile Ranges (IQR) that confidence with the results is established. In case they are not, it might mean that the Likert-style statement did not function properly, for example, respondents may have been confused by the wording; therefore, those responses must be discarded from the dataset and are not reported. There was no such a case in this study.

Since the Likert items and scales produce ordinal data to measure non-numeric concepts and that can be ranked and tallied (Blaikie 2003; Hansen 2003), the statistical approach involves calculating the median as a measure of central tendency, not the mean as it would have been for a probability survey. Moreover, IQRs, not variances or standard deviations, are calculated as the measure of statistical dispersion (Clegg 1998). The interpretation is: the smaller the IQR, the more bunched up the data points around the median; by contrast, the higher the IQR, the more spread out the data points.

Appendix C: The Respondents

The study intends to generate responses that are diverse in geography, focus of activities, membership size, and location of work to achieve rich, diversified responses. Except for the Caribbean and South American regions, all world regions have been represented in the 2014 survey. In a follow-up survey in 2017–2018, only groups from Australia, North America, and Turkey had participated. Majority of the responses in both surveys have come from Australia, Canada, and the USA ($N = 39$ or 87% in 2014 and $N = 7$ or 88% in 2017), indicating that the majority of our respondents work within social spaces in the context of industrialised societies.

Some may criticized the distribution of respondents and the response rate as imbalances. Indeed, only seven of the 39 invited participants opted to respond in our follow up survey. This should not be seen as a limitation. Indeed, the respondent size in a purposively sampled qualitative study—as the case in our surveys—need only be sufficient until the investigators have reached 'the quality of information...rather than the number per se' (Sandelowski 1995: 179). The responses collected for this work, despite their low quantities, appear to have given sufficient variability and richness for them to be considered useful for this book's purposes.

In this study, thirty-four respondents or 72% categorise themselves as environmental groups (in 2017, there are six or 75% of them); two or 4% list their group as primarily oriented as a social justice or human rights group (none in the 2017 cohort identified as such). Eleven or 23% list themselves as 'other,' including: three health-focused groups, two faith or religious groups, one education-focused group, and one petition-oriented group (there are two groups in this category in the 2017 survey). Table C.1 describes the participating social action groups, where our respondents emanate.

Since the respondents' groups are not strictly limited to one particular cause in their focus and work, the respondents were also asked to specify other causes that they are supporting. Forty-one (87%) state that they are also working on climate actions (2017 survey: four or 25%); twelve (26%) are also concerned with health issues (2017 survey: two or 13%); ten (21%) state that they are also focused on the youth sector; eight (17%) are also working on issues about indigenous peoples; seven (15%) are also working on anti-corruption issues; seven also campaign on anti-nuclear and civil rights issues (2017 survey: one for each cause or 6%); five (11%) state that the scope of their work also include anti-racism; three (6%) state that their groups are also working on animal rights; another three (6%) state that anti-globalisation issues are among their other concerns; another three (6%) state that they are also working on labour issues; two (4%) state that LGBT rights are also among their concerns; two (4%) state that direct democracy is a cause they are also working on; another two (4%) state that they work on either women's or education issues. Other causes that the respondents work on include: children and women, decolonisation, environmental law, anti-fracking, livelihoods, parenting, population, refugees, and disability rights.

With regard to coverage and location, 35 respondents (74%) state that they are locally or community-based (2017 survey: three or 21%); twenty-seven (57%) state that they are working at the state or provincial and national levels (2017: 6 or 43%); sixteen (34%) have international reach (2017 survey: only one); and twenty-four (51%) state that they also maintain online presence (2017: two respondents or 14%).

All respondents are involved in education, information, and awareness campaigns through lectures, seminars, trainings, workshops, pamphlet distribution, etc. In addition to education campaigns; 31 or 66% also use non-violent direct action, civil resistance, or civil disobedience such as sit-ins, rallies, demonstrations, strikes, workplace occupations, and/or blockades; 36 or 77% are using both education and non-violent direct action.

Fourteen respondents (30%) state that they have less than 50 members (2017 survey: three or 38%); seven (15%) state that they have between

Table C.1   Social action groups of study respondents

Respondent's code	Description of their social action group
2014 Survey	
SAE01	This U.S.-based climate action group was formed in 2012. The group pushes for climate legislation and policy. It testifies in state legislature, organises petition drives, meets with political candidates, organises communities around their sovereign rights, holds rallies and demonstrations, risks arrest, and meets with industry leaders to discuss concerns. It has several statewide campaigns involving infrastructure resistance and divestment from fossil fuels for endowments and pension funds. It holds monthly meetings in its regional nodes within the state
SAE02	This faith-based Australian group was formed in 2006. It puts up initiatives that promote environmental responsibility in diverse settings from worshipping to school and local communities. It does this through education, advocacy and expressions of ecotheology. The group's policy priorities include sustainability, energy, water and pollution
SAE03	This faith-based Australian group was formed in 2007. It strives to makes its operations more sustainable
SAE04	This U.S.-based environmental group was formed in 2003. It helps individuals with specific environmental problems in their respective neighbourhoods. The group also has taken part in statewide environmental campaigns against fracking
SAE05	This U.S.-based climate action group was formed in 2009. It connects youth environmental activists from across the US Pacific Northwest region to share campaign ideas, brainstorm new tactics, and conduct trainings and workshops
SAE06	This U.S.-based group was formed in 1981. It educates doctors and medical students on the health effects of climate change to spur them to action. It is also involved in local climate change and public health adaptation planning efforts
SAE07	This Hungary-based environmental group was formed in 1988. It primarily works on transport pollution issues, but is also involved with campaigns for greener and cleaner Budapest
SAE08	This Australia-based climate action group was formed in 2006. It undertakes projects aimed at raising climate action awareness including the need for Australian sustainable energy transition
SAE09	This Nepal-based group was formed in 2001. It conducts research-based education and advocacy campaigns, recommends public policy, and undertakes pilot implementation programs on issues related to sustainable energy use and environmental conservation
SAE10	This Canada-based group was formed in 1989. It promotes environmental protection through reduction, reuse, and recycling

(continued)

Table C.1 (continued)

Respondent's code	Description of their social action group
SAE11	This Australia-based climate action group was formed in 2007. It aims to increase its members and communities' climate change awareness and understanding
SAE12	This Australia-based climate action group was formed in 2006. It aims to raise awareness on local, state, and national climate change issues, and environmental and trade exploitation. It was involved in a number of campaigns including surveys, film screenings, market stalls, lobbying a local council, etc.
SAE13	This Australia-based group was founded in 2006. It researches and promotes information about domestic solar photovoltaic rooftop installations. It also lobbies its local council to be more proactively supportive of climate action
SAE14	This US-based group was formed in 2010 to advocate for climate change education
SAE15	This Gambia-based international group was formed in 1994. It works in partnership with local civic groups to support communities to realise their development aspirations. The group works on climate resilient techniques to enhance community livelihoods. It also works on issues about food and nutritional security, resilience to disasters, institutional capacity strengthening, peace building, and governance
SAE16	This Australia-based group was formed in 2008. It provides assistance in teaching sustainability through free-access online learning activities that are linked to the national curriculum
SAE17	This U.S.-based group was formed in 1996. It mobilises citizens against forest destruction and deforestation. It aims to raise awareness about the growing biomass industry. Among its current campaigns are mobilising support to oppose the destruction of the US Southern forests and working with European policy leaders to stop the growing European demand for biomass
SAE18	This Australia-based group was formed in 2008. It engages local residents, businesses, and community groups in its area and surrounding districts to reduce water and energy consumption, minimise waste, and increase biodiversity
SAE19	This Australia-based group was formed in 1996. It is a community legal centre providing free advice on environmental law issues including planning, climate change, environmental assessments, and threatened species protection
SAE20	This Nepal-based group was formed in 2008. It provides capacity building programs for young graduates of natural sciences, researches biodiversity and climate change, appears in media to campaign on environmental awareness, and publishes a weekly newsletter on environmental issues in the Himalayas

(continued)

Table C.1   (continued)

Respondent's code	Description of their social action group
SAE21	This Canada-based group was formed in 2007. It meets once a month for forums or film screenings on environmental issues. It advocates for citywide climate change preparedness. It has been conducting an annual environmental fair that attracts 600–800 people
SAE22	This Australia-based group was formed in 2006. It provides support to its members on low carbon living, and how to lobby politicians on climate action
SAE23	This Turkey-based group was formed in 2004. It works on biodiversity conservation and sustainable natural resources management
SAE24	This Australia based group was formed in 2006. It provides climate action information to its members and the local residents
SAE25	This Canada-based group was formed in 2013. It lobbies local governments of cities and towns to pass laws that would require gasoline retailers to place climate change and air pollution labels on their gas pump nozzles
SAE26	This US-based group was formed in 2010. Its primary work is concerned with clean water and air
SAE27	This Australia-based group was formed in 2011. It develops strategies for triggering a fast build-up of community and governmental commitment to restoring a safe climate through an emergency speed economic mobilisation
SAE28	This India-based group was formed in 1988. It works on local community action for the protection of the environment. The group covers rural areas and focuses on women, children, and the youth
SAE29	This Canada-based group was formed in 1969. It works to protect and conserve British Columbia's wilderness, species and ecosystems within the context of global warming impacts
SAE30	This US-based group was formed in 2011. It works to raise awareness about the Keystone pipeline on local, national and international levels
SAE31	This Australia-based group was formed in 1988. It focuses its work on ecologically sustainable population, both nationally and internationally
SAE32	This South Africa-based group was formed in 2011. It works to resist fracking activities through education and awareness campaigns
SAE33	This Australia-based group was formed in 2007. It engages with the community and people in positions of influence to encourage them to take stronger climate action
SAE34	This US-based group was formed in 2007. It works to empower the younger generation who are most at risk by the climate crisis to be heard by the generation with the most power to enact change. The group trains youth leaders to become spokespersons for their generation in boardrooms, classrooms, courtrooms, the media, Congress, and communities

(continued)

Table C.1   (continued)

Respondent's code	Description of their social action group
SAE35	This US-based group was formed in 2013. The group works online on a petition site that seeks signatures in which signers publicly acknowledge climate change as a threat to civilization and calls on the US Federal Government to instigate a World War 2 scale mobilization. The petition calls for a commitment to reach zero net carbon emissions by 2025 and to devote national resources to mitigate the damage to be caused by climate change in future decades
SAE36	This US-based group was formed in 2012. It calls on Columbia University to divest its endowments from fossil fuel companies
SAE37	This US-based group was formed in 2012. It holds forums, participates in demonstrations, and lobbies for change
SAE38	This Australia-based group was formed in 2008. It works to inform the community about climate change science and relevant issues, lobby governments and representatives to take action, support renewable energy campaigns, support divestment from fossil fuels, and support activities that reduce the impact of anthropogenic climate change
SAE39	This US-based group was formed in 2011. It calls on the University of California to divest its holdings from fossil fuel companies and reinvest these funds in sustainable alternatives
SAJ01	This faith-based Australian group was formed in 2012. It works and dialogs with local communities and their leaders
SAO01	This Canada-based group was formed in 1994. It works on issues around human health and the environment such as climate change and chemical pesticides
SAO02	This US-based group was formed in 1998. It runs a blog and a petition site
SAO03	This Australia-based group was formed in 2010. It has a variety of work including: (1) advocacy for state, national and international policy; (2) secondary research; (3) publishing; (4) communications, education and outreach; (5) community and health sector engagement; (6) developing a network of health service providers engaged in sustainability; (7) running events; (8) liaising with environmental campaigners; (9) online campaigns; (10) offering webinars
SAO04	This US-based group was formed in 2010. It defends the teaching of science in public schools specifically on evolution and climate change
SAO05	This US-based group was formed in 1981. It works on issues surrounding resilience, climate change mitigation, and adaptation. The group has also been working to: (1) eliminate nuclear weapons; (2) support clean, safe renewable energy; (3) eliminate coal and nuclear power; (4) address the health impacts of climate change and of coal and nuclear radioisotope exposure

(continued)

Table C.1   (continued)

Respondent's code	Description of their social action group
SAO06	This Australia-based group was formed in 1957. It is involved in grassroots level organising working most particularly in anti-nuclear action
SAO07	This UK-based group was formed in 1968. This faith-based international development agency works on long-term development and disaster relief
2017 Survey	
SAE40	This U.S.-based group was formed in 2013. It works in the Berkshire
SAE41	This Australia-based group was formed in 2011. It develops strategy for emergency speed restoration of a safe climate
SAE42	This US-based group was formed in 2008. It educates, advocates and offers workshops, forums and conferences and community-based trainings on climate resiliency, nuclear disarmament advocacy, promotion of clean, safe, renewable energy and urging regulation of environmental toxins
SAE43	This Australia-based group was formed in 2013. It connects investors with owners of commercial properties needing finance for solar installations
SAO08	This Australia-based group was formed in 2010. It does advocacy, research, policy and communications to develop policy guidance on health and climate change for governments and other groups. This group participated in the 2014 survey and was coded as SAO03
SAE44	This Canada-based group was formed in 1985. It conducts public education at events and in schools
SAO09	This US-based group was formed in 1981. It defends the integrity of science education
SAE45	This Turkey-based group was formed in 2004. It works on biodiversity research, conservation and sustainable use of natural resources. This group participated in the 2014 survey and was coded as SAE23

51 and 100 members (2017 survey: two or 25%); five (11%) say that they have between 101 and 250 members (nil during the 2017 survey); two (4%) have between 251 and 500 members (2017 survey: two or 25%); six (13%) have membership size from 501 to 999; six others (13%) have membership size between 1,000 and 4,999; seven (15%) state that they have more than 5,000 members (2017 survey: one or 12%).

REFERENCES

Ackerman, P., & Kruegler, C. (1994). *Strategic Nonviolent Conflict: The Dynamics of People Power in the Twentieth Century.* Westport, CT: Praeger.

Adam, E. M. (2017). Intersectional coalitions: The paradoxes of rights-based movement building in LBTQ and immigrant communities. *Law & Society Review, 51,* 132–167.

Agyeman, J. (2008). Toward a 'just' sustainability? *Continuum: Journal of Media & Cultural Studies, 22,* 751–756.

Agyeman, J., Doppelt, B., Lynn, K., & Hatic, H. (2007). The climate-justice link: Communicating risk with low-income and minority audiences. In S. C. Moser & L. Dilling (Eds.), *Creating a Climate for Change: Communicating Climate Change and Facilitating Social Change* (pp. 119–138). Cambridge, UK: Cambridge University Press.

American Psychological Association (APA). (2009). *Psychology and Global Climate Change: Addressing a Multi-Faceted Phenomenon and Set of Challenges: A Report by the Task Force on the Interface Between Psychology and Global Climate Change.* Washington, DC: APA.

Anderson, K., & Nevins, J. (2016). Planting seeds so something bigger might emerge: The Paris Agreement and the fight against climate change. *Socialism and Democracy, 30,* 209–218.

Andrews, K. T., Ganz, M., Baggetta, M., Han, H., & Lim, C. (2010). Leadership, membership, and voice: Civic association that work. *American Journal of Sociology, 115,* 1191–1242.

Arabella Advisors. (2014). *Measuring the Global Fossil Fuel Divestment Movement.* http://bit.ly/1reDL6Q.

© The Editor(s) (if applicable) and The Author(s), under exclusive licence to Springer International Publishing AG, part of Springer Nature 2019
L. L. Delina, *Climate Actions,*
https://doi.org/10.1007/978-3-319-91884-6

Assadourian, E. (2013). Building an enduring environmental movement. In E. Assadourian & T. Prugh (Project Directors), *State of the World 2013: Is Sustainability Still Possible?* (pp. 292–303). Washington, DC; Covelo, CA; and London, UK: Island Press.

Australian Youth Climate Coalition (AYCC). (2013). *About AYCC.* http://bit.ly/1pm8daW.

Bacon, W. (2013). *Climate Science in Australian Newspapers.* Sydney, Australia: The Australian Centre for Independent Journalism.

Bacon, W., & Nash, C. (2012). Playing the media game: The relative (in)visibility of coal industry interests in media reporting of coal as a climate change issue in Australia. *Journalism Studies, 13,* 243–258.

Barboza, T. (2014, September 22). L.A., Houston, Philadelphia mayors vow more action on climate change. *Los Angeles Times.* http://lat.ms/1odyfOn.

Bertini, I. (2014, September 23). Church of Sweden completed fossil fuels divestment as movement doubles in one year. *Blue & Green Tomorrow.* http://bit.ly/1wLlKyM.

Beveridge, R., & Kern, K. (2013). Energiewende in Germany: Background, developments and future challenges. *Renewable Energy Law Policy Review, 4,* 3–12.

Billings, P. (2013, December 8). Protesters in Hodgman's sights. *The Examiner.* http://bit.ly/1d5a931.

Blaikie, N. (2003). *Analysing quantitative data.* London, UK: Sage Publications.

Bomberg, E., & McEwen, N. (2012). Mobilizing community energy. *Energy Policy, 51,* 435–444.

Bond, J. (2001). The media and the movement: Looking back from the southern front. In B. Ward (Ed.), *Media, Culture, and the Modern African American Freedom Struggle* (pp. 16–40). Gainesville: University Press of Florida.

Bostrom, A., Böhm, G., & O'Connor, R. E. (2013). Targeting and tailoring climate change communications. *WIREs Climate Change, 4,* 447–455.

Boykoff, M. T., & Boykoff, J. M. (2004). Balance as bias: Global warming and the US prestige press. *Global Environmental Change, 14,* 125–136.

Boykoff, M. T., & Goodman, M. K. (2009). Conspicuous redemption? Reflections on the promises and perils of the 'celebritization' of climate Change. *Geoforum, 40,* 395–406.

Brown, L. R., & Betsill, M. (2008). *Plan B 3.0: Mobilizing to Save Civilization.* New York, London, UK, and Washington, DC: W. W. Norton.

Bulkeley, H., & Betsill, M. (2003). *Cities and Climate Change: Urban Sustainability and Global Governance.* London, UK: Routledge.

Buys, L., Aird, R., Van Megen, K., Miller, E., & Sommerfeld, J. (2014). Perceptions of climate change and trust in information providers in rural Australia. *Public Understanding of Science, 23,* 170–188.

C40 Cities Climate Leadership Group (C40). (2015). *C40 Cities.* http://bit.ly/1ir0TJ5.

C40, UN Secretary General's Special Envoy for Cities and Climate Change, & Stockholm Environment Institute. (2014, September). *Advancing Climate Ambition: Cities as Partners in Global Climate Action.* http://bit. ly/1yGBKXC.

C40, United Cities and Local Governments, & International Council for Local Governments. (n.d). *The Compact of Mayors.* http://bit.ly/1pFz9ni.

CAN International. (n.d.). About Us. http://www.climatenetwork.org.

Carson, L. (2010, September 9). Growing up politically: Conducting a national conversation on climate change. *Australian Policy Online.* http://bit.ly/ ONNRaH.

Carson, L. (2011). Dilemmas, disasters and deliberative democracy: Getting the public back into policy. *Griffith Review Edition 32: Wicked Problems, Exquisite Dilemmas, 32*(Winter), 25–32.

Center for Research on Environmental Decisions (CRED). (2009). *The Psychology of Climate Change Communication: A Guide for Scientists, Journalists, Educators, Political Aides, and the Interested Public.* New York, NY: CRED, Columbia University.

Chandler, D. (2004). Building global civil society 'from below'? *Millennium: Journal of International Studies, 33,* 313–339.

Chenoweth, E., & Stephan, M. J. (2011). *Why Civil Resistance Works: The Strategic Logic of Nonviolent Conflict.* New York: Columbia University Press.

Chomsky, N. (2012). *Occupy.* New York: Zuccotti Park Press.

Cialdini, R. B. (1993). *Influence: The Psychology of Persuasion* (2nd Rev. ed.). New York: Quill-William Morrow.

Cinderby, S., Haq, G., Cambridge, H., & Lock, K. (2015). Building community resilience: Can everyone enjoy a good life? *Local Environment, 21,* 1252–1270.

City Government of New York. (2014). *One City, Built to Last: Transforming New York City's Buildings for a Low-Carbon Future.* http://on.nyc. gov/1pJQQRB.

City of Eugene. (2010). *A Community Climate and Energy Action Plan for Eugene.* http://bit.ly/1vEpT6r.

City of Evanston. (2012, May 3). Evanston to Buy Its Own Energy, Save Residents Money and Help Achieve Climate Action Goals. http://bit.ly/16Nw8aa.

City of Sydney. (2013). *City of Sydney Decentralised Energy Master Plan Renewable Energy 2012–2030.* http://bit.ly/1f0DWx3.

Cleantechnica. (2013). *Germany: 100% Renewable Energy and Beyond.* RenewEconomy. http://bit.ly/12e7iT6.

Clegg, F. (1998). *Simple Statistics.* Cambridge, UK: Cambridge University Press.

Climate Action Tracker. (2015, November 13). *G20—All INDCs in, but Large Gap Remains.* Berlin, Germany: Climate Analytics and Ecofys; Potsdam, Germany: Potsdam Institute for Climate Impact Research; and Cologne, Germany: NewClimate Institute.

Climate Reality Project. (n.d.). *Climate Reality Leadership Corps*. http://bit. ly/1yU5wFv.

Collins, R. (2010). The contentious social interactionism of Charles Tilly. *Social Psychology Quarterly, 73*, 5–10.

Cumbers, A., Routledge, P., & Nativel, C. (2008). The entangled geographies of global justice networks. *Progress in Human Geography, 32*, 183–201.

Cvetkovich, G., & Löfstedt, R. (Eds.). (2000). *Social Trust and the Management of Risk: Advances in Social Science Theory and Research*. London, UK: Earthscan.

Daly, H., & Farley, J. (2004). *Ecological Economics: Principles and Applications*. Washington, DC: Island Press.

Dean Moore, K., & Nelson, M. P. (2013). Moving toward a global moral consensus on environmental action. In *State of the World 2013: Is Sustainability Still Possible?* Washington, DC, Covelo, and London, UK: Island Press.

Delina, L. (2016). *Strategies for Rapid Climate Mitigation: Wartime Mobilisation as Model for Action?* Oxon, UK: Routledge.

Delina, L. (2018). Can energy democracy thrive in a non-democracy? *Frontiers in Environmental Science, 6*, 5.

Delina, L., & Diesendorf, M. (2013). Is wartime mobilisation a suitable policy model for rapid national climate mitigation? *Energy Policy, 58*, 371–380.

Delina, L., Diesendorf, M., & Merson, J. (2014). Strengthening the climate action movement strategies from histories. *Carbon Management, 5*, 397–409.

Della Porta, D. (2005). Multiple belongings, tolerant identities, and the construction of 'another politics': Between the European social forum and the local social fora. In D. Della Porta & S. Tarrow (Eds.), *Transnational Protest and Global Activism*. Lanham, MD: Rowman & Littlefield.

Della Porta, D., & Diani, M. (2006). *Social Movements: An Introduction* (2nd ed.). Malden, MA: Blackwell.

Della Porta, D., Andretta, M., Mosca, L., & Reiter, H. (2006). *Globalization from Below*. London, UK: University of Minnesota Press.

Democracy Now! (2014, September 21). *Special 3-Hour Broadcast of the People's Climate March*. http://bit.ly/1tHsEFF.

Diani, M. (1992). The concept of social movement. *The Sociological Review, 401*, 1–25.

Diani, M. (1995). *Green Networks*. Edinburgh, UK: Edinburgh University Press.

Diesendorf, M. (2009). *Climate Action: A Campaign Manual for Greenhouse Solutions*. Sydney, Australia: UNSW Press.

Dietz, R., & O'Neill, D. (2013). *Enough Is Enough: Building a Sustainable Economy in a World of Finite Resources*. San Francisco, CA: Berrett-Koehler.

Dryzek, J. (2000). *Deliberative Democracy and Beyond: Liberals Critics, Contestations*. Oxford, UK: Oxford University Press.

Dryzek, J. (2009). Democratization as deliberative capacity building. *Comparative Political Studies, 42,* 1379–1402.

Dunlap, R. E., & McCright, A. M. (2011). Organized climate change denial. In J. S. Dryzek, R. B. Norgaard, & D. Schlosberg (Eds.), *The Oxford Handbook of Climate Change and Society* (pp. 144–160). Oxford, UK: Oxford University Press.

Ereaut, G., & Segnit, N. (2006). *Warm words: How are we telling the climate story and can we tell it better?* London, UK: Institute for Public Policy Research.

Evans, J. P. (2011). Resilience, ecology and adaptation in the experimental city. *Transactions of the Institute of British Geographers, 36,* 360–376.

Farrell, J. (2011). *Democratizing the Electricity System: A Vision for the 21st Century Grid.* http://bit.ly/1eZrYH8.

Featherstone, D. (2005). Towards the relational construction of militant particularisms: On why the geographies of past struggles matter for resistance to neoliberal globalisation. *Antipode, 37,* 250–271.

Featherstone, D. (2008). *Resistance, Space and Political Identities: The Making of Counter-Global Networks.* Oxford, UK: Wiley-Blackwell.

Featherstone, H., Weitkamp, E., Ling, K., & Burnett, F. (2009). Defining issue-based publics for public engagement: Climate change as a case study. *Public Understanding of Science, 18,* 94–101.

Feldman, L., Maibach, E. W., Roser-Renouf, C., & Leiserowitz, A. (2011). Climate on cable: The nature and impact of global warming coverage on Fox News, CNN, and MSNBC. *The International Journal of Press/Politics, 17,* 3–32.

Fernandez, R. M., & MacAdam, D. (1988). Social networks and social movements: Multiorganizational fields and recruitment to Mississippi Freedom Summer. *Sociological Forum, 3,* 357–382.

Fessenden-Raden, J., Fitchen, J. M., & Heath, J. S. (1987). Providing risk information in communities: Factors influencing what is heard and accepted. *Science, Technology, and Human Values, 12,* 94–101.

Fisher, D. R. (1998). Rumoring theory and the internet: A framework for analysing the grass roots. *Social Science Computer Review, 16,* 158–168.

Fisher, D. R., Dow, D. M., & Ray, R. (2017). Intersectionality takes it to the streets: Mobilizing across diverse interests for the Women's March. *Science Advances, 3,* eaao1390.

Fisher, D. R., & McInerney, P.-B. (2012). The limits of networks in social movement retention: On canvassers and their careers. *Mobilization: An International Journal, 17,* 109–128.

Floyd, D. L., Prentice-Dunn, S., & Rogers, R. W. (2000). A meta-analysis of research on protection motivation theory. *Journal of Applied Social Psychology, 30,* 407–429.

Foderaro, L. W. (2014, September 21). Taking a call for climate change to the streets. *The New York Times.* http://nyti.ms/1qkVZzy.

Foran, J. (2014). "¡Volveremos!/we will return": The state of play for the global climate justice movement. *Interface, 6,* 454–477.

Foxon, T. J. (2013). Transition pathways for a UK low carbon electricity future. *Energy Policy, 52,* 10–24.

FrameWorks Institute. (2002). *Framing Public Issue.* Washington, DC: FrameWorks Institute.

Fung, A. (2008). Democratizing the policy process. In R. Goodin, M. Rein, & M. Moran (Eds.), *The Oxford Handbook of Public Policy* (pp. 669–688). Oxford, UK: Oxford University Press.

Ganz, M. (2004). Why David sometimes wins: Strategic capacity in social movements. In D. M. Messick & R. M. Kramer (Eds.), *The Psychology of Leadership: New Perspectives and Research.* Malwah, NJ: Lawrence Erlbaum Associates, Publishers.

Ganz, M. (2006a). Strategy, deliberation and meetings. In *Organizing Course Notes.* Cambridge, MA: Harvard Kennedy School. http://bit.ly/1Dx2ppv.

Ganz, M. (2006b). Mobilizing power: Analysis, strategy, deliberation. In *Organizing Course Notes.* Cambridge, MA: Harvard Kennedy School. http://bit.ly/1xcbrGW.

Ganz, M. (2010). *Why David Sometimes Wins: Leadership, Organization, and Strategy in the California Farm Worker Movement.* Oxford, UK: Oxford University Press.

Garland, R. (1991). The mid-point on a rating scale: Is it desirable? *Marketing Bulletin, 2,* 66–70.

Gifford, R. (2011). The dragons of inaction: Psychological barriers that limit climate change mitigation and adaptation. *American Psychology, 66,* 290–302.

Gilding, P. (2011). *The Great Disruption: How the Climate Crisis Will Transform the Global Economy.* London, UK: Bloomsbury.

Giugni, M., & Grasso, M. T. (2015). Environmental movements in advanced industrial democracies: Heterogeneity, transformation and institutionalization. *Annual Review of Environment and Resources, 40,* 337–361.

Gleditsch, K. S., & Celestino, M. R. (2013). Fresh carnations or all thorn, no rose? Non-violent campaigns and transitions in autocracies. *Journal of Peace Research, 50,* 385–400.

Go 100% Renewable Energy. (n.d.). *About Us.* http://bit.ly/1umA6U1.

Gudynas, E. (2011). Buen Vivir: Germinando alternativas al desarrollo. *América Latina en movimiento, 462,* 1–20.

Gutmann, A., & Thompson, D. (2004). *Why Deliberative Democracy?* Princeton, NJ: Princeton University Press.

Haas, T., & Sander, H. (2016). Shortcomings and perspectives of the German Energiewende. *Socialism and Democracy, 30,* 121–143.

Haidt, J. (2007). The new synthesis on moral psychology. *Science, 316,* 998–1002.

Hall, J. D. (2005). The long civil rights movement and the political uses of the past. *Journal of American History, 91,* 1233–1263.

Hall, N., & Taplin, R. (2008). Room for climate advocates in a coal-focused economy? NGO influence on Australian climate policy. *Australian Journal of Social Issues, 43,* 359–379.

Hall, N., Taplin, R., & Goldstein, W. (2010). Empowerment of individuals and realization of community agency: Applying action research to climate change responses in Australia. *Action Research, 8,* 71–91.

Hannam, P. (2014, October 20). More Australian universities come under pressure to divest from fossil fuels. *The Sydney Morning Herald.* http://bit.ly/1qLBZI3.

Hansen, J. P. (2003). CAN'T MISS—conquer any number task by making important statistics simple. Part 1. Types of variables, mean, median, variance and standard deviation. *Journal for Healthcare Quality, 25,* 19–24.

Hess, D. J. (2018). Energy democracy and social movements: A multi-coalition perspective on the politics of sustainability transitions. *Energy Research & Social Science, 40,* 177–189.

Hetherington, K. (1997). *Expressions of Identity.* London, UK: Sage.

Hopkins, R. (2008). *The Transition Handbook: From Oil Dependency to Local Resilience.* Totnes, Devon, UK: Green Books.

Houck, D. W., & Dixon, D. E. (Eds.). (2006). *Rhetoric, Religion and the Civil Rights Movement, 1954–1965.* Waco, TX: Baylor University Press.

Intergovernmental Panel on Climate Change (IPCC). (2014). *Climate Change 2014: Synthesis Report of the Fifth Assessment Report of the IPCC.* The Core Writing Team, R. K. Pachauri & L. Meyer (Eds.). Switzerland: IPCC.

IPCC Working Group III. (2014). *Summary for Policymakers in Climate Change 2014: Impacts, Adaptation, and Vulnerability* (Fifth Assessment Report of the IPCC). Switzerland: IPCC.

Isham, J., & Waage, S. (Eds.). (2007). *Ignition: What You Can Do to Fight Global Warming and Spark a Movement.* Washington, DC: Island Press.

Islar, M., & Busch, H. (2016). "We are not in this to save the polar bears!"—The link between community renewable energy development and ecological citizenship. *Innovation: The European Journal of Social Science Research, 29,* 303–319.

Jasanoff, S. (2004). Science and citizenship: A new synergy. *Science and Public Policy, 31,* 90–94.

Jasper, J. M., & Poulsen, J. D. (1995). Recruiting strangers and friends: Moral shocks and social networks in animal rights and antinuclear protests. *Social Problems, 42,* 493–512.

Jenkins, J. C. (1985). *The Politics of Insurgency: The Farm Workers Movement in the 1960s.* New York: Columbia University Press.

Jenkins, K., McCauley, D., Heffron, R., Stephan, H., & Rehner, R. (2016). Energy justice: A conceptual review. *Energy Research & Social Science, 11*, 174–182.

Johnson T. (2014). *Sunday News Shows Ignore Historic Climate March*. Media Matters for America. http://bit.ly/1wWLr27.

Jost, J. T., Napier, J. L., Thorisdottir, H., Gosling, S. D., Palfai, T. P., & Ostafin, B. (2007). Are needs to manage uncertainty and threat associated with political conservatism or ideological extremity? *Personality and Social Psychology Bulletin, 33*, 989–1007.

Kahan, D. M., & Braman, D. (2008). The self-defensive cognition of self-defense. *American Criminal Law Review, 45*, 1–65.

Kahan, D. M., Jenkins-Smith, H., & Braman, D. (2011). Cultural cognition of scientific consensus. *Journal of Risk Research, 14*, 147–174.

Karagiannopoulos, V. (2012). The role of the internet in political struggles: Some conclusions from Iran and Egypt. *New Political Science, 34*, 151–171.

Kirby, A. (2008). *A UN Guide to Climate Neutrality*. Malta: United Nations Environment Programme and GRID-Arendal.

Klandermans, B., & Oegema, D. (1987). Potentials, networks, motivations, and barriers: Steps towards participation in social movements. *American Sociological Review, 52*, 519–531.

Klein, N. (2014). *This Changes Everything: Capitalism vs. the Climate*. New York: Simon & Schuster.

Krosnick, J. (2013). *Public Opinion on Global Warming in Texas: 2013*. Stanford, CA: Woods Institute for the Environment, Stanford University.

Krosnick, J. A., & MacInnis, B. (2013). Does the American public support legislation to reduce greenhouse gas emissions? *Daedalus, the Journal of the American Academy of Arts & Science, 142*, 26–39.

Kurzman, C. (1996). Structural opportunity and perceived opportunity in social movement theory: The Iranian Revolution of 1979. *American Sociological Review, 61*, 153–170.

Kurzman, C. (2012). The Arab Spring: Ideals of the Iranian Green Movement, methods of the Iranian Revolution. *International Journal of Middle East Studies, 44*, 162–165.

Læssoe, J. (2007). Participation and sustainable development: The post-ecologist transformation of citizen involvement in Denmark. *Environmental Politics, 16*, 231–250.

Landy, D. (2015). Bringing the outside field interaction and transformation from below in political struggles. *Social Movement Studies, 14*, 255–269.

Largest Climate March in History—Your Pictures. (2014, September 22). *The Guardian*. http://bit.ly/1uRQI94.

Leiserowitz, A., Maibach, E., & Roser-Renouf, C. (2009). *Global Warming's 'Six Americas' 2009: An Audience Segmentation Analysis*. New Haven, CT: Yale Project on Climate Change: School of Forestry and Environmental Studies, Yale University; Fairfax, VA: Center for Climate Change Communication and George Mason University.

Leviston, Z., Price, J., Malkin, S., & McCrea, R. (2014). *Fourth Annual Survey of Australian Attitudes to Climate Change: Interim Report.* Canberra: CSIRO.

Lorenzoni, I., Nicholson-Cole, S., & Whitmarsh, L. (2007). Barriers perceived to engaging with climate change among the UK public and their policy implications. *Global Environmental Change, 17,* 445–459.

Lovell, H., Hann, V., & Watson, P. (2018). Rural laboratories and experiments at the fringes: A case study of a smart grid on Bruny Island, Australia. *Energy Research & Social Science, 36,* 146–155.

MacLeod, J. (2012). *Civil Resistance in West Papua (Perlawanan tanpa kekerasan di Tanah Papua).* Ph.D. thesis, School of Political Science and International Studies, University of Queensland, Brisbane, QLD, Australia.

Mann, M. (1993). *The Sources of Social Power, Volume 2: The Rise of Classes and Nation-States, 1760–1914.* Cambridge, UK: Cambridge University Press.

Manning, S., & Reinecke, J. (2016). A modular governance architecture in-the-making: How transnational standard-setters govern sustainability transitions. *Research Policy, 45,* 618–633.

Mansbridge, J., Bohman, J., Chambers, S., Christiano, T., Fung, A., et al. (2012). A systemic approach to deliberative democracy. In J. Parkinson & J. Mansbridge (Eds.), *Deliberative Systems: Deliberative Democracy at the Large Scale* (pp. 1–26). Cambridge, UK: Cambridge University Press.

Mark, J. (2013). Conversation: Naomi Klein. *Earth Island Journal* (Autumn). http://bit.ly/18R9oZ2.

Markowitz, E. M., & Shariff, A. F. (2012). Climate change and moral judgment. *Nature Climate Change, 2,* 243–247.

Martin, B. (2010). Theory for activists. *Social Anarchism, 44,* 22–41.

Marx, G. T. (1967). Opiate or inspiration of civil rights militancy among Negroes? *American Sociological Review, 32,* 64–72.

Marx, S. M., Weber, E. U., Orlove, B. S., Leiserowitz, A., Krantz, D. H., Roncoli, C., et al. (2007). Communication and mental processes: Experiential and analytic processing of uncertain climate information. *Global Environmental Change, 17,* 47–48.

Maxmin, C. (2017, June 5). How Harvard Divestment Was Won. *The Nation.*

McAdam, D. (1999). *Political Process and the Development of Black Insurgency, 1930–1970.* Chicago, IL: University of Chicago Press.

McAdam, D. (2011). The US civil rights movement: Power from below and above, 1945–70. In A. Roberts & T. G. Ash (Eds.), *Civil Resistance and Power Politics: The Experience of Non-violent Action from Gandhi to the Present.* Oxford, UK: Oxford University Press.

McCauley, D., Rehner, R. W. M., & Pavlenko, M. (2015). Assessing the justice implications of energy infrastructural development in the Arctic. In R. J. Heffron & G. Little (Eds.), *Delivering Energy Law and Policy in the EU and US.* Edinburgh, UK: Edinburgh University Press.

McKibben, B. (2012, July 19). Global warming's terrifying new math. *Rolling Stone.* http://rol.st/LuRoru.

McKibben, B. (2013). Beyond baby steps: Analysing the cap-and-trade flop. *Grist.* http://bit.ly/1dGK3sU.

McKibben, B. (2013, April 11). The fossil fuel resistance. *Rolling Stone.*

McNamara, M. (2014, January 29). Criminalising activism a growing trend. *Echo Net Daily.* http://bit.ly/100GuYE.

Melucci, A. (1989). *Nomads of the Present.* Philadelphia: Temple University Press.

Melucci, A. (1996). *Challenging Codes.* Cambridge, UK: Cambridge University Press.

Millar, M. (2013, November 19). Harper government's extensive spying on anti-oil sands group revealed in FOIs. *Vancouver Observer.* http://Bit.Ly/1ddnsqs.

Mirkinson, J. (2014, September 22). TV news misses yet another opportunity to cover climate change. *The Huffington Post.* http://huff.to/1sUOqkX.

Monbiot, G. (2006). *Heat: How to Stop the Planet Burning.* London: Allen Lane.

Mooney, C. (2011, May/June). The science of why we don't believe science. *Mother Jones.* http://bit.ly/2Ez9egt.

Moore, K. D., & Nelson, M. P. (Eds.). (2010). *Moral Ground: Ethical Action for a Planet in Peril.* San Antonio, TX: Trinity University Press.

Morris, C., & Jungjohann, A. (2016). *Energy Democracy: Germany's Energiewende to Renewables.* London, UK: Palgrave Macmillan.

Moser, S. C. (2007a). Communication strategies to mobilize the climate movement. In J. Isham & S. Waage (Eds.), *Ignition: What You Can Do to Fight Global Warming and Spark a Movement* (pp. 73–95). Washington, DC: Island Press.

Moser, S. C. (2007b). In the long shadows of inaction: The quiet building of a climate protection movement in the United States. *Global Environmental Politics, 7,* 124–144.

Moser, S. C. (2007c). More bad news: The risk of neglecting emotional responses to climate change information. In S. C. Moser & L. Dilling (Eds.), *Creating a Climate for Change: Communicating Climate Change and Facilitating Social Change* (pp. 64–80). Cambridge, UK: Cambridge University Press.

Moser, S. C. (2009). Costly knowledge—Unaffordable denial: The politics of public understanding and engagement on climate change. In M. T. Boykoff (Ed.), *The Politics of Climate Change: A Survey* (pp. 155–181). Oxford, UK: Routledge.

Moser, S. C., & Dilling, L. (Eds.). (2007). *Creating a climate for change: Communicating climate change and facilitating social change.* Cambridge, UK: Cambridge University Press.

Moyer, B. (1987). *The Movement Action Plan: A Strategic Framework Describing the Eight Stages of Successful Social Movements.* The Social Movement Empowerment Project. http://bit.ly/1f9KI6p.

Moyer, B., McAllister, J., Finley, M. L., & Soifer, S. (2001). *Doing Democracy: The MAP Model for Organizing Social Movements.* Gabriola Island, BC: New Society Publishers.

Mungiu-Pippidi, A., & Munteanu, I. (2009). Moldova's "Twitter revolution". *Journal of Democracy, 20,* 136–142.

Murphy, G. (2005). Coalitions and the development of the global environmental movement: A double-edged sword. *Mobilization, 10,* 235–250.

Nagda, B. R. A. (2006). Breaking barriers, crossing borders, building bridges: Communication processes in intergroup dialogues. *Journal of Social Issues, 62,* 553–576.

National Research Council. (2002). *New Tools for Environmental Protection: Education, Information, and Voluntary Measures.* Washington, DC: National Academy Press.

National Science Foundation (NSF). (2009). *Science and Engineering Indicators 2006.* Arlington, VA: NSF.

Nature (editorial). (2010). Climate of Fear. *Nature, 464,* 141.

Nepstad, S. E. (2011). *Nonviolent Revolutions: Civil Resistance in the Late 20th Century.* New York: Oxford University Press.

New Economics Foundation (NEF). (2008). *A Green New Deal: Joined-Up Policies to Solve the Triple Crunch of the Credit Crisis, Climate Change and High Oil Prices.* London, UK: NEF.

Niemeyer, S., & Jennstål, J. (2018). *Scaling up deliberative effects—Applying lessons of minipublics.* Oxford, UK: Oxford University Press Handbook of Deliberative Democracy, Oxford University Press.

Nisbet, M. C. (2009). Communicating climate change: Why frames matter for public engagement. *Environment, 51,* 12–23.

North, P. (2011). The politics of climate activism in the UK: A social movement analysis. *Environment and Planning A, 43,* 1581–1598.

North, P., & Longhurst, N. (2013). Grassroots localization? The scalar potential of and limits of the 'transition' approach to climate change and resource constraint. *Urban Studies, 50,* 1423–1438.

Nunes, R. (2009). What were you wrong about ten years ago? *Turbulence, 5,* 38–39.

O'Neill, S. J., & Hulme, M. (2009). An iconic approach for representing climate change. *Global Environmental Change, 19,* 402–410.

O'Neill, S. J., & Nicholson-Cole, S. (2009). "Fear won't do it": Promoting positive engagement with climate change through visual and iconic representations. *Science Communication, 30,* 355–379.

Oettingen, G. (2015). *Rethinking Positive Thinking: Inside the New Science of Motivation*. New York: Penguin.

Opp, K.-D., & Gern, C. (1993). Dissident groups, personal networks, and spontaneous cooperation: The East-German revolution of 1989. *American Sociological Review, 58,* 659–680.

Orr, D. W. (2013). Governance in the long emergency. In E. Assadourian & T. Prugh (Eds.), *State of the World 2013: Is Sustainability Still Possible?* (pp. 279–291). Washington, DC; Covelo; and London, UK: Island Press.

Oteman, M., Wiering, M., & Helderman, J.-K. (2014). The institutional space of community initiatives for renewable energy: A comparative case study of the Netherlands, Germany and Denmark. *Energy Sustainability and Society, 4,* 11.

Parag, Y., Capstick, S., & Poortinga, W. (2011). Policy attribute framing: A comparison between three policy instruments for personal emissions reduction. *Journal of Policy Analysis and Management, 30,* 889–905.

Parag, Y., & Janda, K. B. (2014). More than filler: Middle actors and socio-technical change in the energy system from the middle-out. *Energy Research & Social Science, 3,* 102–112.

Parag, Y., Hamilton, J., White, V., & Hogan, B. (2014). Network approach for local and community governance of energy: The case of Oxfordshire. *Energy Policy, 62,* 1064–1077.

Parliament of Victoria. (2013). *Summary Offences and Sentencing Amendment Bill 2013*. Victoria, Australia. http://bit.ly/P8nAtd.

Parliament of Victoria. (2014). *Summary Offences and Sentencing Amendment Act 2014*. Victoria, Australia. http://bit.ly/1rkMbvW.

Pearse, R. (2016). Moving targets: Carbon pricing, energy markets, and social movements in Australia. *Environmental Politics, 25,* 1079–1101.

Pennicuik, S. (2014, March 11). *On the Summary Offences and Sentencing Amendment Bill 2013*. Website of the Victorian Greens. http://bit.ly/1mA33x6.

People's Climate March. (2014). *Wrap Up*. http://bit.ly/1qxNEZt.

Plumer, B. (2013, January 16). *Why Has Climate Legislation Failed?* An interview with Theda Skocpol. *The Washington Post.* http://wapo.st/1jFMBUA.

Polleta, F., & Jasper, J. M. (2001). Collective identity and social movements. *Annual Review of Sociology, 27,* 283–305.

Raushenbush, P. B. (2013, July 2). Fossil fuel divestment strategy passes at United Church of Christ Convention (UCC). *The Huffington Post.* http://huff.to/1xVzYfC.

Renewables 100 Policy Institute. (2007). *What We Are*. http://bit.ly/15AqFL8.

Rhodes, R. A. W. (1996). The new governance: Governing without government. *Political Studies, 64,* 652–667.

Rosewarne, S., Goodman, J., & Pearse, R. (2014). *Climate Action Upsurge: The Ethnography of Climate Movement Politics*. Abingdon, Oxfordshire, UK, and New York: Routledge.

Routledge, P. (2012). Translocal climate justice solidarities. In J. S. Dryzek, R. B. Norgaard, & D. Schlosberg (Eds.), *The Oxford Handbook of Climate Change and Society*. Oxford, UK: Oxford University Press.

Ruskin, G. (2013). *Spooky Business: Corporate Espionage Against Nonprofit Organizations*. http://bit.ly/1DGhUeQ.

Sandelowski, M. (1995). Sample size in qualitative research. *Research in Nursing & Health, 18*, 179–183.

Saunders, C. (2008). Double-edged swords? Collective identity and solidarity in the environmental movement. *The British Journal of Sociology, 59*, 227–253.

Saunders, C. (2009). It's not just structural: Social movements are not homogeneous responses to structural features, but networks shaped by organisational strategies and status. *Sociological Research Online, 14*, 1–16.

Saunders, C. (2013). Insiders, thresholders, and outsiders in west European global justice networks: Network position and modes of coordination. *European Political Science Review, 2*, 167–189.

Scheufele, D. A., Corley, E. A., Dunwoody, S., Hih, T. J., Hillback, A., & Guston, D. H. (2007). Scientists worry about some risks more than the public. *Nature Nanotechnology, 4*, 91–94.

Schlosberg, D., & Dryzek, J. S. (2002). Digital democracy: Authentic or virtual? *Organization & Environment, 15*, 332–335.

Schock, K. (2005). *Unarmed Insurrections: People Power Movement in Nondemocracies*. Minneapolis: University of Minnesota Press.

Schneider, F., Kallis, G., & Martinez-Alier, J. (2010). Crisis or opportunity? Economic degrowth for social equity and ecological sustainability. *Journal of Cleaner Production, 18*, 511–518.

Schor, J. B. (2010). *Plenitude: The New Economics of True Wealth*. New York: The Penguin Press.

Segnit, N., & Ereaut, G. (2007). *Warm Words II: How the Climate Story Is Evolving and the Lessons We Can Learn for Encouraging Public Action*. London, UK: Institute for Public Policy Research.

Seyfang, G., Park, J. J., & Smith, A. (2013). A thousand flowers blooming? An examination of community energy in the UK. *Energy Policy, 61*, 977–989.

Sharp, G. (1973a). *The Politics of Nonviolent Action: Part One, Power and Struggle*. Boston, MA: Porter Sargent.

Sharp, G. (1973b). *The Politics of Nonviolent Action: Part Two, the Method of Nonviolent Action*. Boston, MA: Porter Sargent.

Sharp, G. (2005). *Waging Nonviolent Struggle: 20th Century Practice and 21st Century Potential*. Boston, MA: Porter Sargent.

Shirky, C. (2011). The political power of social media: Technology, the public sphere and political change. *Foreign Affairs, 90*, 28–41.

Skocpol, T. (2013). *Naming the Problem*. Cambridge, MA: Harvard University Symposium on the Politics of America's Fight Against Global Warming. http://bit.ly/Pc8W3O.

Slaughter, A.-M. (2005). *A New World Order*. Princeton, NJ: Princeton University Press.

Smith, N., & Leiserowitz, A. (2013). The role of emotion in global warming support and opposition. *Risk Analysis, 34*, 937–948.

Snow, D. A. (2013). Framing and social movements. In D. A. Snow, D. della Porta, B. Klandermans, & D. McAdam (Eds.), *The Wiley-Blackwell Encyclopedia of Social and Political Movements*. Malden, MA: Blackwell.

Snow, D. A., & Benford, R. D. (1988). Ideology, frame resonance, and participant mobilization. *International Social Movement Research, 1*, 197–217.

Snow, D. A., Zurcher, L. A., & Ekland-Olson, S. (1980). Social networks and social movements: A microstructural approach to differential recruitment. *American Sociological Review, 45*, 787–801.

Snow, D. A., Rochford, B., Worden, S. K., & Benford, R. D. (1986). Frame alignment processes, micromobilization, and movement participation. *American Sociological Review, 51*, 464–481.

Sovacool, B. K., & Dworkin, M. H. (2014). *Global Energy Justice: Principles Problems and Practices*. Cambridge, UK: Cambridge University Press.

Stepanova, E. (2011). *The Role of Information Communication Technologies in the "Arab Spring": Implications Beyond the Region*. Washington, DC: PONARS Eurasia, the George Washington University, and Elliott School of International Affairs. http://bit.ly/1L9ceOg.

Stern, P. C. (2012). Psychology: Fear and hope in climate messages. *Nature Climate Change, 2*, 572–573.

Stirling, A. (2008). 'Opening up' and 'closing down': Power, participation and pluralism in the social appraisal of technology. *Science, Technology, and Human Values, 33*, 262–294.

Stoknes, P. E. (2015). *What We Think About When We Try Not to Think About Global Warming*. White River Junction, VT: Chelsea.

Strathman, A., Gleicher, F., Boninger, D. S., & Edwards, C. S. (1994). The consideration of future consequences: Weighing immediate and distant outcomes of behaviour. *Journal of Personality and Social Psychology, 66*, 742–752.

Suarez, S. L. (2006). Mobile democracy: Text messages, voter turnout and the 2004 Spanish general election. *Representation, 42*, 117–128.

Sullivan, C. (2010, December 14). San Francisco eyes goal of 100% green power by 2020. *The New York Times*. http://nyti.ms/P8KPU7.

Swim, J. K., Stern, P. C., Doherty, T. J., Clayton, S., Reser, J. P., Weber, E. U., et al. (2011). Psychology's contributions to understanding and addressing global climate change. *American Psychologist, 66*, 241–250.

Tarrow, S. (1998). *Power in Movement: Social Movements and Contentious Politics*. New York: Cambridge University Press.

Tarrow, S. (2005). *The New Transnational Activism*. Cambridge, UK: Cambridge University Press.

Thomas, T. (2013, December 19). A room full of eco-idiots. *The Daily Telegraph*. http://bit.ly/1uJcBFk.

Tilly, C. (1978). *From Mobilization to Revolution*. Reading, MA: Addison-Wesley.

Tilly, C. (1995). *Popular Contention in Great Britain, 1754–1834*. Cambridge, MA: Harvard University Press.

Tilly, C. (2002). *Stories, Identities and Political Change*. New York: Rowman & Littlefield.

Tilly, C. (2006). *Regimes and Repertoires*. Chicago, IL: University of Chicago Press.

Tilly, C. (2008). *Contentious Performances*. Cambridge, UK: Cambridge University Press.

UNFCCC. (2015). *Adoption of the Paris Agreement* (Report FCCC/CP/2015/L.9/Rev.1). http://unfccc.int/resource/docs/2015/cop21/eng/l09r01.pdf.

United Nations. (1992). *United Nations Framework Convention on Climate Change* (FCCC/INFORMAL/84). http://unfccc.int/resource/docs/convkp/conveng.pdf.

United Nations Human Settlements Programme (UN-Habitat). (2011). *Hot Cities: Battle-Ground for Climate Change*. Nairobi, Kenya: UN-Habitat.

Vala, C. T., & O'Brien, K. J. (2007). Attraction without networks: Recruiting strangers to unregistered protestantism in China. *Mobilization, 12*, 79–94.

Van der Schoor, T., & Scholtens, B. (2015). Power to the people: Local community initiatives and the transition to sustainable energy. *Renewable and Sustainable Energy Reviews, 43*, 666–675.

Van der Schoor, T., Van Lente, H., Scholtens, B., & Peine, A. (2016). Challenging obduracy: How local communities transform the energy system. *Energy Research & Social Science, 13*, 94–105.

Wagner, G., & Zeckhauser, R. J. (2012). Climate policy: Hard problem, soft thinking. *Climatic Change, 110*, 507–521.

Wardekker, J. A., Petersena, A. C., & Sluijs, J. P. (2009). Ethics and public perception of climate change: Exploring the Christian voices in the US public debate. *Global Environmental Change, 19*, 512–521.

Weber, T. (1997). *On the Salt March*. New Delhi, India: HarperCollins.

Weber, E. U. (2006). Experience-based and description-based perceptions of long-term risk: Why global warming does not scare us (yet). *Climatic Change, 77*, 103–120.

Weber, E. U. (2010). What shapes perceptions of climate change? *WIREs Climate Change, 1*, 332–342.

Wolcott, J. (2007, May 1). Rush to judgment. *Vanity Fair*. http://vnty.fr/1lJ80yE.

World Bank. (2012). *Turn Down the Heat: Why a 4 °C Warmer World Must Be Avoided*. Washington, DC: The World Bank.

Yagatich, W., Galli Robertson, A. M., & Fisher, D. R. (2018). How local environmental stewardship diversifies democracy. *Local Environment, 23*, 431–447.

Index

119